Our Good Earth

OUR GOOD EARTH

A NATURAL HISTORY OF SOIL

BERMAN D. HUDSON, PH.D.

Algora Publishing
New York

Library of Congress Cataloging-in-Publication Data

Names: Hudson, Berman D., author.
Title: Our good earth : a natural history of soil / Berman D. Hudson.
Description: New York : Algora Publishing, [2020] | Includes
    bibliographical references and index. | Summary: "The author explains
    the science and the importance of soil, with a description of how soils
    have evolved over the past 3.5 billion years and how they affect human
    civilization"— Provided by publisher.
Identifiers: LCCN 2020000576 (print) | LCCN 2020000577 (ebook) | ISBN
    9781628943962 (hardcover) | ISBN 9781628943955 (trade paperback) | ISBN
    9781628943979 (adobe pdf)
Subjects: LCSH: Soils.
Classification: LCC S591 .H843 2020 (print) | LCC S591 (ebook) | DDC
    631.4—dc23
LC record available at https://lccn.loc.gov/2020000576
LC ebook record available at https://lccn.loc.gov/2020000577

Printed in the United States

Acknowledgment

I would like to thank Dr. Pattie West for her editorial guidance and moral support throughout the writing of this book.

Table of Contents

CHAPTER 1. WHAT IS SOIL AND WHY DOES IT MATTER?

> The dust lifted up out of the fields and drove gray plumes into the air... the finest dust did not settle back to earth, but disappeared into the darkening sky...the corn fought the wind with its weakened leaves until the roots were freed by the prying wind and then each stalk settled wearily sideways...and the wind cried and whimpered over the fallen corn.

The words above are from John Steinbeck's 1939 novel *The Grapes of Wrath*, for which he won both the National Book Award and the Pulitzer Prize. In this acclaimed work, Steinbeck gave a human face to the Dust Bowl, one of the worst environmental disasters in American history. The Dust Bowl is the name once given to a vast region of nearly 100 million acres that includes sections of Kansas, Oklahoma, Colorado, New Mexico, and Texas. For thousands of years most of this area had been covered in grass, but beginning in the 1800s, land-hungry farmers spread out across the region and soon the grass was gone; and with the coming of drought early in the 20th century, the soil began to blow away.

The "dusters," as the storms came to be called, began in the early 1930s after a long period of unusually high temperatures. The weather service reported 179 dust storms in 1933 alone and in 1935 a single storm destroyed five million acres of wheat in Kansas, Oklahoma, and Nebraska. Large storms carried clouds as far east as New York City. For three consecutive years, an average of nine storms a month hit Amarillo, Texas, during the four-month period from January through April. Houses, barns, farm equipment, and fields were covered in layers of dust more than 20 feet thick.

Even after the storms ended, so much dust would be suspended in the air that the sun remained invisible for several days. A woman from Garden City, Kansas, describes one such storm: "The doors and windows were all shut tight, yet those tiny particles seemed to seep through the very walls. It got into cupboards and clothes closets; our faces were as dirty as if we had rolled in the dirt; our hair was gray and stiff; and we ground dirt between our teeth." The storms continued for almost a decade, wreaking havoc on the regional economy and exacting an enormous toll in human misery. Farms and ranches failed, forcing thousands of desperate families to load up their belongings and migrate west in search of better lives (Editors, Time-Life Books 1985).

Plowing up a vast area of grassland and exposing the bare soil to strong winds in a region with sporadic rainfall was a huge mistake, but fortunately it was done in a wealthy country with a surplus of land. America has an abundance of land and a relatively small population, but the world as a whole is not so fortunate. In 1900 there were 1.6 billion people on Earth. Within only a century, world population had nearly quadrupled, reaching 6.1 billion, and demographers working for the United Nations estimate that there will be 11.0 to 12.0 billion people in the world by 2100. Some think the UN estimates are too conservative and that world population will be even higher than their 11.2 billion projection.

In 1900, there were a little more than 2.0 hectares of arable land for each person on Earth; by 2100 this will have declined to around 0.3 hectare per person. Arable means land that currently is being used to grow crops or could be used for that purpose in the future. If you have trouble visualizing how big a hectare is, just remember that an American football field is about one half hectare in size. We lose some arable land every year, but that is not the real problem; instead, as the world's population grows larger, each person's share keeps getting smaller. Table 1.1 shows how rapidly the amount of arable land per person has declined recently as a result of population growth.

Here is how the figures in Table 1.1 were derived. A team of Dutch scientists recently estimated that the world has about 3.42 billion hectares of arable land (Buringh 1977), a figure in close agreement with an earlier assessment made by the President's Science Advisory Committee (PSAC 1967). In 1900 there were 1.6 billion people on Earth. Dividing 3.42 billion hectares by 1.6 billion people gives us 2.1 hectares per person, the figure in the table above. By 2000, world population had grown to 6.1 billion, but the global supply of arable land had stayed about the same. Dividing 3.42

billion hectares by 6.1 billion people gives us 0.6 hectare per person, the figure in the table for the year 2000.

Table 1.1. Hectares of arable land in the world per person, 1800 to 2100

Year	Arable land per person (hectares)
1800	3.4
1900	2.1
2000	0.6
2100 (projected)	0.3

In 2100, there will be about 0.30 hectare of arable land per capita. This means that each person, on average, will have to survive on the food produced by a parcel of land about half the size of an American football field — not just for a time, but for his or her entire life, all year long, in winter and summer, in good years and bad. This is assuming that world population peaks at 11.2 billion as predicted, which is far from certain. It also assumes that that the amount of arable land lost to such things as erosion, salinization, rising seas, and urbanization will be kept to a minimum, but that too is uncertain. The most likely scenario is that population will be higher than predicted at the end of this century and the total amount of arable land will have declined.

The numbers above highlight the fact that conserving the world's soil and managing it well are critical issues for the 21st century and beyond. With population continuing to surge in much of the world, we soon could be very close to the Malthusian edge. Each year, as more and more people are born, the margin for error gets smaller. In order to safeguard the world's soil, we first must recognize it for the irreplaceable resource that it is. Second, we must understand it better. For example, which soils are the most fertile, which are the most fragile, which require the most inputs? In addition to agriculture, we must recognize the other critical roles that soil plays in the economy of nature. In summary, we must learn more about soil and teach more of what we have learned to more people.

What Is Soil?

Scientists have struggled with this question for more than a century. Nearly everyone has a general idea of what soil is, but coming up with a scientific definition has been surprisingly hard. Hans Jenny, in his classic

1941 text, *Factors of Soil Formation*, wrote, "In the layman's mind, the soil is a very concrete thing, namely the 'dirt' on the surface of the earth. To the soil scientist, or pedologist, the word 'soil' conveys a somewhat different meaning, but no generally accepted definition exists." Jenny then presented some definitions that had been put forth early in the 20th century by a number of highly regarded European and American soil scientists. He did not think any of their definitions were entirely satisfactory, but Jenny did not propose one himself, instead proclaiming, "It is problematic whether any definition of soil could be formulated to which everyone would agree. Fortunately, there is no urgent need for universal agreement. For the purposes of presentation and discussion of the subject matter it is necessary only that the reader know what the author has in mind when he [or she] uses the word 'soil'."

Charles Kellogg, an early leader of the American soil survey program, expressed similar sentiments a decade later in a review article published in *Scientific American* (Kellogg 1950). In this article, Kellogg suggested that pedologists might actually be offended if someone asked them to define soil. In Kellogg's words, they might perceive the question as "awkwardly vague — like asking what is rock, or mineral, or gas." He then suggested that a pedologist might respond to such an impertinent question in the following way: "There are thousands of kinds of soils, each as individual in character and properties as a species of rock, of plant, or of animal. Every soil has its own unique history; like a living organism it is the creature of a dynamic process of evolution, and a revealing subject for scientific study."

Despite the inherent difficulty of defining soil, the Soil Science Society of America (SSSA) has had an official definition for some years. Considering who they are, there is almost no way they could have avoided the issue. The SSSA recently revised its definition, with a committee working for two years before presenting a draft version to the Society's governing board in the spring of 2017. The following definition was approved a few months later (van Es 2017).

> Soil: The layer(s) of generally loose mineral and/or organic material that are affected by physical, chemical, and/or biological processes at or near the planetary surface and usually hold liquids, gases, and biota and support plants.

There are two important points to be made about this definition. First, it refers to soil as a "material" as nearly everyone attempting to define soil has done for more than a century. This is accurate as far as it goes, but it does not go far enough. Defining soil as a material provides little real information, since almost anything that has mass can be described as such. Little is specified, for example, about the nature and properties of this material or materials. The definition then states that soils usually "hold liquids, gases, and biota and support plants." This also is a true statement, but it does not get to the essence of the matter; it does not identify the properties of the soil itself that enable it to do these things. The annotated definition below provides more specifics and, I believe, gives us greater insight into the true nature of soil.

> Soil: Layers consisting of mineral particles 2.0 millimeters or less in diameter and/or highly decomposed or fibrous organic material that form on land surfaces where rocks, air, water, and living organisms come together and interact. A soil can be entirely mineral, entirely organic, or a combination of the two. Whether mineral or organic, the defining characteristics of soil are *porosity, capillarity, and cation exchange.*

Porosity refers to the fact that one third to one half or more of soil volume is empty space. Of course, the so-called empty space is never really empty; the larger pores and voids are filled with air, while the smaller ones hold water against the downward pull of gravity. Because of porosity, the roots of vascular plants can find a foothold in soil and grow in all directions in search of water and nutrients. Porosity also makes it possible for the soil to provide water, air, and living space to the multitude of small life forms that find a home there.

Capillarity, the second defining attribute of soil, is the capacity of soils to take in water and hold it against the downward pull of gravity. The water stored in small soil pores serves as a reserve from which plant roots can make withdrawals on a continuing basis. If soil pores did not store water and if they were not refilled partially or fully after each rain, terrestrial life would be much impoverished and human agriculture as we know it would not exist.

Cation exchange, the third defining attribute of soil, results from negative charges on the surfaces of small mineral particles (mostly clay) and organic matter. Since nature does not tolerate charge imbalances, these negative charges attract and hold positively charged ions such as Ca^{2+}, Mg^{2+}, and K^+ to even things out. Attached ions are exchangeable and serve as a nutrient supply for plant roots and soil microorganisms. Cation exchange in soil is one of the most important phenomena in nature. If this capacity did not exist, many familiar ecosystems would be missing from our world. Without the electrical charges on clay and organic matter to hold them in place, elements such as calcium, magnesium, and potassium would be leached away quickly, greatly limiting the production of biomass on land; and without cation exchange in soil, agriculture and human civilization as we know it would not be possible.

It is important to emphasize that porosity, capillarity, and cation exchange are the defining attributes of soil; if all three are not present, then it is not soil. This helps to clarify something that might be confusing to some people — the idea that soil can be totally mineral or totally organic. This seems counterintuitive. How can two different entities, one derived from weathered rock and another consisting of highly decomposed or fibrous organic matter, be the same thing; how can both be called soil? Organic soils and mineral soils can be put in the same category because, despite their extreme differences in composition, both form on land surfaces and both exhibit the three attributes that define soil: porosity, capillarity and cation exchange.

A Posteriori Knowledge

The definition above clearly differentiates soil from everything else in the universe, but there is a problem. Like nearly all definitions, it assumes some degree of what Immanuel Kant (1781) referred to as a posteriori knowledge, which is knowledge gained from prior experience. The problem of a posteriori or prior knowledge can best be explained by using a simple example. Carbon is commonly defined as a chemical element in which each atom has six protons. This is a good definition because nothing else in the universe meets this criterion. However, the definition of carbon, like that of soil, assumes some level of prior knowledge. Most people reading this book can understand the definition of carbon because somewhere, perhaps in high school or college, they learned what an element is, what an atom is, and what a proton is; but without such previously acquired knowledge, the definition would be meaningless to them.

Soil is more complex than a single element such as carbon, so defining it is much harder, and a definition of soil will, of necessity, assume a lot more prior knowledge. Considering this, defining soil near the beginning of this book might seem unwise, but it was done for a reason. A detailed definition is presented early in the narrative to serve as a template or guide — to highlight the knowledge and concepts we must acquire in order to fully understand soil and how it forms. This brings up an interesting question: Is it possible to understand how such a complex entity as soil is formed and then convey that understanding to others?

Francis Bacon and Modern Science

Francis Bacon published his great philosophical work *Novum Organum* (the New Method) in 1620 as a direct challenge to Aristotle's *Organum* written more than 16 centuries earlier. Bacon championed the use of observation, measurement and inductive reasoning to understand the world. This is in contrast to Aristotle's syllogistic method that had dominated Western thinking for centuries. Syllogism is a logical method by which one uses deductive reasoning to arrive at a conclusion (i.e., knowledge) by starting out with two or more propositions that are assumed to be true. Nearly everyone has seen an example of this process, such as "All men are mortal. Aristotle is a man. Therefore, Aristotle is mortal." Bacon argued that very little real knowledge can be gained in this way; instead, one must use inductive reasoning; examine nature itself, gather facts, and then use these facts to discern the laws and principles that govern the world.

The empiricism espoused by Bacon has been the guiding principle of science for the past 400 years, enabling such visionaries as Johannes Kepler, Isaac Newton, James Clerk Maxwell, and Albert Einstein to gradually make the mysteries of nature known to us. We now know that the universe came into being about 13.7 billion years ago, and 400 years of scientific inquiry have taught us a lot about how it works. Although random events happen, we now know that the universe overall is orderly, predictable, and is governed by laws and principles that can be perceived and understood by the human mind.

But how many such laws and principles are there? According to a 2009 book by Robert Hazen and James Trefil, there are relatively few of them. Hazen, a geologist, and Trefil, a physicist, came up with just 19 big ideas or central concepts that, in their view, one must understand in order to be scientifically literate. Some examples: The universe is regular and predictable; all matter is made of atoms; stars experience a cycle of birth and death; all life is based on the same chemistry and genetic code. Being scientifically literate does not mean that you know everything there is to know; it simply means that you have acquainted yourself with the common principles and facts that underlie all fields of science.

The approach Hazen and Trefil took is an insightful one and can be of great value in understanding things that are heterogeneous and complex, of which soil is a striking example. Accordingly, this book follows their lead, but with a much narrower focus, posing a much narrower question: "What are the core concepts of science needed to understand soil — how it forms, what it does, and why there are so many kinds of it in the world?" Listed below are the eight central concepts of science that seem most relevant to these questions. The list below combines some items from Hazen and Trefil's list and expands or adds detail to others. In addition, the list includes one concept that was not on their list; this is the idea that there are four fundamental forces of nature: the strong nuclear force, the weak nuclear force, gravity and electromagnetism. The interplay between gravity and electromagnetism is especially important to the formation and functioning of soil. Here is the list of scientific principles most relevant to the understanding of soil.

- Although random events happen, the universe overall is regular, predictable, and can be understood by the human mind.

- Four fundamental forces govern the universe: the weak nuclear force, the strong nuclear force, gravity, and electromagnetism.

- All matter is made up of atoms, most of which can combine with other atoms by sharing electrons.

- The behavior of matter depends on how its atoms are arranged and the nature of the bonds holding them together.

- The Earth's land surface is constantly changing as a result of tectonic forces from below and weathering agents from above.

- Heat results from the rapid movement or agitation of atoms. Heat moves spontaneously from a hot body to a cold body by convection, conduction, or radiation.

- All life forms on Earth evolved through natural selection and use the same basic chemistry and the same genetic code.

- Life persists generation after generation by recycling the same water and the same chemical elements over and over again.

Anyone with a working knowledge of these eight concepts will have little trouble understanding how soils form, how they function, how they vary across the world, and how they evolved. If you took introductory chemistry and biology in high school or college, you probably have the prerequisite knowledge, but such prior knowledge is not necessary. These are simple, commonsense concepts and each is explained in the text as it becomes relevant to the narrative. If you have been exposed to the information before, this will be a good review; if not, it will serve as an introduction. Using the eight central concepts of science outlined above as touchstones, this book explores five major themes or topics related to soil:

1. How soils form
2. What soils do
3. How soils vary spatially
4. How soils evolved
5. Soil and agriculture

Soil is composed of mineral matter, organic matter, air, water, and living organisms. The air and water in soil are little affected by the processes of soil formation. Soil air usually contains more carbon dioxide and less oxygen than free air, but it has the same general composition. The same is true of water; it can exist in soil as a liquid, a gas, or a solid (ice), the same as it does outside the soil. But the mineral and organic matter that make up soil are different; both have been greatly altered by the processes of weathering and soil formation. Mineral particles in soil are classified according to size as sand, silt, or clay. Most of the larger particles, sand and silt, result directly from the physical and chemical breakdown of rocks, but clay does not. If you analyze a piece of granite, you will not find any clay, but granitic rocks weather to produce clayey soils. Clay is manufactured by recombining some of the elements that are released as the rock weathers; the formation of mineral soil involves both the breakdown of rocks and the synthesis of clay.

The rocks at the Earth's surface are made up almost entirely (98.8 percent) of just eight elements — oxygen, silicon, aluminum, iron, calcium, magnesium, sodium, and potassium. Chapter 2 describes how, as the Earth cooled, these elements made their way to the surface and came to dominate the near-surface rocks. Understanding rock weathering and soil formation requires that we understand how these eight elements are combined to form rocks and what happens to each of them when rock is dissolved by weathering processes. "Dissolved" might seem like a strange word to use here, but rocks do dissolve, much as a sugar cube does in a cup of tea, albeit much more slowly. Chapter 3 discusses the chemical makeup of rocks in detail and explains why some elements are more susceptible than others to being removed from the rock structure and leached away. Chapter 4 describes how clay forms in soil and explains why certain elements are preferentially held back to be used in its formation. Chapter 5 focuses largely on sodium and calcium, the most mobile of the rock forming elements, explaining why the atoms of both elements are carried away to the sea in such large quantities — and how many of these atoms make their way back to land. The details of their individual journeys and the importance of each to terrestrial life make for interesting narratives.

Organic Matter

Unlike mineral matter, which is derived from the breakdown of rocks, soil organic matter is manufactured mostly from air and water via photosynthesis. Carbon dioxide from the air and water from the soil are the raw materials of this process, while the sun supplies the energy, as shown by the formula below.

$$6 \, CO_2 \, (\text{air}) + 6 \, H_2O \, (\text{soil}) + \text{energy from the sun} \Rightarrow C_6 H_{12} O_6 + 6 O_2$$

During photosynthesis, water molecules are split and the oxygen is given off to the atmosphere, while the hydrogen is kept. The hydrogen is then combined with carbon dioxide from the air to form glucose or $C_6 H_{12} O_6$. This basic sugar molecule is then used as a source of energy and as a raw material to form a wide variety of the compounds that make life possible. Most of these compounds, though much larger and much more complex than the basic glucose molecule, still consist entirely or mostly of carbon, hydrogen, and oxygen. Examples include cellulose, lignin, chitin, fats, starches, and proteins. Since carbon, hydrogen, and oxygen dominate the basic molecules of life, it follows that living organisms overall are made up mostly of these three elements. This includes people. By weight, more than 90 percent of the human body consists of carbon, hydrogen, and oxygen.

Humans are not fashioned from the clays of the Earth. Instead, our bodies consist mostly of two elements taken from air (carbon and oxygen) and one element derived from water (hydrogen) — and we are not unique in this respect. Every living thing on Earth is comprised mostly of these three elements. Since every bit of organic matter in soil was once part of a living organism, it follows that soil organic matter too is made up mostly of carbon, hydrogen, and oxygen. We therefore can draw two important distinctions between soil mineral matter and soil organic matter. First, they differ as to their origins. Soil mineral matter is derived from the breakdown of rocks, whereas soil organic matter is derived ultimately from photosynthesis. Second, since soil organic matter is derived from living organisms, it is carbon-based, whereas the mineral matter is silicon-based.

Carbon is just above silicon in the same column of the periodic table, so the two elements have much in common. Both have four electrons in their outermost shell. In order to achieve chemical stability, both elements must give away or share those four electrons. A significant difference between the two elements is the way they achieve this. Chapter 6 highlights the similarities and differences between carbon and silicon and addresses some fundamental questions about them. For example, why is life on Earth carbon-based, while silicon is of little use to living things but is integral to the structure of rocks and minerals?

Soil Organisms

The fifth component of soil is the organisms that live there. These include fungi and bacteria, as well as a diverse assortment of invertebrates such as earthworms and soil-dwelling insects, all of which depend directly or indirectly on soil organic matter for their sustenance. Each year, green plants remove about 120 gigatons of carbon from the air. The prefix giga means one billion, or 10^9, so a gigaton is one billion metric tons. Since the total amount of carbon in the atmosphere is only about 750 gigatons, green plants remove about 16 percent of the world supply each year (120/750). This would be a problem were it not for soil; but fortunately, while plants are busy growing above ground, soil organisms are hard at work both below ground and on the soil surface. Every bit of organic matter that falls to the ground is soon attacked by a multitude of organisms, broken down into very fine parts, and thoroughly mixed into the soil. The ultimate products of this decomposition are CO_2 and H_2O. Worldwide, this process essentially replaces the carbon that is taken out of the air and stored in new plant tissue each year. Because of the continuous cycling of CO_2, the amount of carbon in soil, air, and biomass remains fairly constant over time. Chapters 6 and 8 present more detailed information on the role of soil organisms in the cycling of carbon and other critical elements.

How Soils Vary Spatially

Soils cover most of the world's land surface and as environmental factors vary across this surface, so do soils; and they do so in a very predictable fashion. Jenny (1941) examined the nature of this variation in detail and cited numerous examples to show conclusively that five factors — climate (cl), organisms (o), relief (r), parent material (p), and time (t) are the "independent variables that define the soil system." He wrote, "for a given combination of cl, o, r, p, and t, the state of the soil system is fixed; only one type of soil exists under these conditions." In essence, the soil at any location on Earth is a function of climate, organisms, relief, and time, or in symbolic form: $s = f\,(cl, o, r, p, t)$.

- Climate (cl) — Climate includes such things as how hot or cold the area is, how much rain it receives, and how rain is distributed throughout the year.

- Organisms (o) — This usually refers to the native vegetation of the area. Even if a soil is now used to grow crops, many of its properties developed during the time it was covered in native vegetation such as forest, grassland, or savanna.

- Relief or topography (r) — Where does the soil occur on the landscape? Is it on a ridge, on a side slope, or in a depression; is it flat, steep, or gently sloping?

- Parent material (p) — This refers to the kind of material from which the soil formed. For example, did it form from granite, from basalt, from limestone, from recently deposited alluvium, or from organic matter?

- Time (t) — How long has the soil been forming? Was the parent material deposited recently or has it been in place and stable for a very long time?

The soil factor equation represents a simple but very powerful concept — so powerful and so useful that it bears repeating: For a given combination of climate, organisms, relief, parent material, and time, only one type of soil exists. Assume you go out into the field, examine an area that has a certain combination of (cl, o, r, p, t) and find that it has a certain kind of soil. If you go a mile away, or ten miles away, 100 miles away, or even 1,000 miles away and find an area with the same combination of cl, o, r, p, and t, then you can expect to find the same kind of soil. This means that the soil factor model or equation can be used to explain how soils vary across the landscape and to make precise, accurate maps showing their distribution.

But this requires that we have identifiable kinds of soil to study or to map, which brings us to Soil Taxonomy, the official system for classifying soils in the United States. An important feature of Soil Taxonomy is its distinctive hierarchical structure. At the top, all of the world's soils are placed into twelve broad groups called soil orders. Each order is then broken down into increasingly narrow subdivisions: orders > suborders > great groups > subgroups > families > series. The orders of Soil Taxonomy represent the twelve major kinds of soil in the world and are a useful category for discussing how soils form and how they vary across the landscape. Assume that an astronaut is orbiting the Earth at a distance of 20,000 miles. Assume further that someone has cleared away the atmosphere and

somehow delineated the world's major soil regions on the globe. Our astronaut could see these delineations from space, and each of these delineations would be dominated by one of the twelve orders recognized in Soil
Taxonomy.

Let us further assume that our astronaut is looking down on North
America. He or she would see one large soil region occupying the desert
areas of the American southwest. Soil Taxonomy assigns the dominant
soils in this region to the order Aridisols, which are soils that are dry most
of the year and cannot be used to grow crops without irrigation. Now let
us assume that our astronaut focuses in on the northeastern part of United
States, and more specifically on an area that includes northeastern New
York and the New England states. Spodosols are the dominant soils in this
region; these are the ashy gray, highly leached soils that commonly form
in rainy climates under coniferous vegetation. Spodosols are dominant
in the northeastern United States because the combinations of climate,
organisms, relief, parent material, and time (cl, o, r, p, t) that occur there
lead to their formation. Similarly, Aridisols are common in the American
southwest because the combinations of (cl, o, r, p, t) that occur there favor
the formation of Aridisols. You will not find Spodosols in the deserts of
Arizona and you will not find Aridisols forming beneath coniferous forests
of Maine or New Hampshire.

These are but two examples; different soil orders are predominant in
geographic areas that have the particular combinations of cl, o, r, p, and t
that favor their formation. The environmental factors leading to the formation of each soil order are discussed at length in Chapters 7 and 8. Below
is a list of the twelve soil orders, which represent the major kinds of soil
existing in the world. You will notice that the name of each order ends
with the suffix sols. You also will notice that the orders are clustered in
groups based on when they first appeared on Earth.

3.8 to 3.5 billion years ago
- Entisols — Soils with little or no profile development
- Inceptisols — Soils that have some profile development, but lack any
  of the horizons or conditions that define the other orders
- Andisols — Relatively young soils, mostly of volcanic origin, that are
  characterized by minerals with poorly organized crystalline structure

2.5 to 1.5 billion years ago
- Gelisols — Very cold soils with permafrost in the subsoil

- Vertisols — Very clayey soils that shrink and crack when dry and expand when wet

- Oxisols — Highly weathered, naturally infertile soils of tropical regions

- Aridisols — The dry soils of deserts

470 to 360 million years ago

- Histosols — Soils that form in decaying organic matter, usually in low, wet areas

- Spodosols — Highly leached, infertile soils that have accumulations of organic matter and iron and aluminum oxides in the subsoil

- Alfisols — Naturally fertile, high-base soils with a clay-enriched subsoil horizon

- Ultisols — Naturally infertile, low-base soils with a clay-enriched subsoil horizon

65 to 50 million years ago

- Mollisols — Very dark colored, naturally fertile soils of grasslands

How Soils Evolved

Chapter 9 discusses the timeline above in detail and explains how and why each soil order evolved. Only three of the existing soil orders — Entisols, Inceptisols, and Andisols, were present prior to 3.0 billion years ago. Interestingly, another soil evolved at the time, but it no longer exists. A mysterious "Green Clay" appeared sometime before 3.0 billion years ago and persisted for more than a billion years before going extinct. At the time these Green Clays formed, there was no free oxygen in the atmosphere. When green plants started pumping oxygen into the air millions of years later, these soils were doomed, since they could not form in an oxygenated atmosphere. As far as we know, this is the only major soil group to evolve on Earth, persist for a long period of time, and then go extinct.

It is noteworthy that the soil orders did not appear one at a time nor all at once; instead, for the most part they arrived in groups, with long intervals of time between the groups. This is a classic example of punctuated equilibrium, a concept introduced by Eldridge and Gould in 1972. In their opinion, much of evolution can be characterized as a process in which long periods of stasis or equilibrium are interrupted at intervals by onsets of rapid environmental change during which new species arise and some old species die out. After each flurry of activity, a new stasis or equilibrium is established. The new equilibrium might last a very long time, but eventually something will disrupt the balance once again, setting the stage for yet another burst of evolutionary activity. Soil evolution is a remarkably good example of this process.

Chapter 9 shows how useful the soil factor equation is in explaining how soils have evolved over time. Prior to 3.0 billion years ago, only Entisols, Inceptisols, and Andisols existed because the available combinations of climate, relief, and parent material existing on Earth at that time could lead only to those three kinds of soils. But during the next 3.0 billion years, the evolution of the lithosphere, biosphere, and atmosphere resulted in new kinds of climate, new kinds of topography, new kinds of organisms, and new kinds of parent material, making it possible for nine additional soil orders to appear. Major precipitating events include the formation of large continents, oxygenation of the atmosphere, the formation of deserts, glaciation, colonization of the land by vascular plants, and the evolution of roots. Chapter 9 explains the significance of each precipitating event and how it led to the evolution of one or more new soil orders.

The Nitrogen Problem

Friedrich Albert Fallou was a German lawyer who specialized in legal matters relating to the appraisal and taxation of land. Becoming concerned over what he saw as the decline of soil quality in his region, he abandoned the practice of law and devoted all of his time to the study of soil. He published a book in 1862 titled *Pedology or General and Special Soil Science*, arguing that pedology, the study of soil, should be recognized as an independent scientific discipline. Although he eventually was overshadowed by Dokuchaev in Russia and Hilgard in America, it can be argued that Fallou is the true founder of soil science. In his writings, he was careful to distinguish between what he called pédologie, the scientific study of soil for its own sake, and agrologie, the practical study of soil as it relates to agriculture.

This book focuses mostly on pédologie, although some issues related to agrologie have crept into a few chapters. However, the last chapter of the book, Chapter 10, deals explicitly with matters related to agriculture, including a consideration of why Jared Diamond (1987) famously referred to agriculture as "the worst mistake in the history of the human race." But most of Chapter 10 is devoted to discussing how the shortage of a single chemical element, nitrogen, limited agricultural production and, as a result, the size and health of human populations for thousands of years. In essence, it focuses on what has been described as the "perverse chemistry of nitrogen."

The Invisible Soil

There are some unusual barriers both to understanding soil and to teaching others about it. One of these is its near invisibility — the fact that soil can rarely be seen because it is hidden away underground. We might catch a glimpse of it in a plowed field, a fresh road cut, or on a construction site. But we rarely see soil in its entirety, and even if we catch sight of an entire soil profile, it lacks the visual impact of plants, animals, and other more photogenic features of the environment. Soil is essential to human survival and if people do not really see it and rarely think about it, they are unlikely to understand just how important it is and why protecting it is one of humanity's most pressing issues. The near invisibility of soil also is a problem for those writing books about it. How can people relate to a subject if they cannot form a clear mental image of what it is the author is writing about?

Whenever soil scientists need to study a soil profile in detail, they dig a pit, which is laborious if done by hand and expensive if done by machinery. Fortunately, hundreds of soil profile pictures have been catalogued and put on the internet for public access. One of the best resources is the State Soils Website maintained by the Natural Resources Conservation Service (NRCS). Keying a phrase such as "NRCS state soils" into a search engine will get you to the website. Once there, scroll down to the table labeled Representative and State Soils and click on the fact sheet for Harney, the state soil of Kansas. This brings up a page with a brief description of Harney and photos of a typical profile and the typical setting or landscape where the soil occurs. This is an especially useful website. It includes a representative soil for each of the 50 states plus Puerto Rico and the Virgin Islands, representing eleven of the twelve soil orders. Histosols are the only soil order not represented. Soil science faculties at several universities also maintain soil-related websites, with the one currently maintained by the University of Idaho deserving special mention. Because of the soil "invisibility" problem, this book relies heavily on internet sources, especially images, to help readers get a better sense of what soil really is.

Soils of Other Worlds

Before ending this chapter, a brief discussion of soil formation on other planetary bodies is in order. Many of the surficial deposits on the Moon and on Mars meet the criteria listed in Soil Taxonomy for mineral soil. For example, at least part of the Moon's surface appears to be covered by loose, fine-grained material consisting of silt-size mineral particles interspersed with rocks and boulders. Moon soil is estimated to be more than 15 meters thick in some areas. The Moon has no water, no atmosphere, and no life. So far, in fact, there is no evidence of life anywhere in the universe other than on Earth. As one might expect in a waterless, airless, sterile environment such as the Moon, soil formation has been largely a physical process, mostly due to the pulverizing effects of meteoroid and micrometeoroid bombardment. Meteoroids large and small have been striking the Moon for several billion years, so it is not surprising that the lunar surface has been so shattered and pulverized, sometimes to a depth of many meters (Eglinton et al. 1972).

Mars, unlike the Moon, does have an atmosphere, but it is a thin one and its composition is very different from that of Earth. The Martian atmosphere is 95 percent carbon dioxide, three percent nitrogen and two percent argon, with only traces of oxygen and water vapor. Several lines of evidence suggest that Mars once had a much denser atmosphere and a lot of surface water, but most of the atmosphere has been lost to space and the surface has been extremely arid for the past two billion years. Some have suggested that many of the soils on Mars are relict; that they formed before the planet lost most of its water and atmosphere, and then were preserved by freeze drying as the planet grew colder and more arid. Data from the Viking probes of the 1970s revealed Martian soils that are shallow and only slightly weathered, reminiscent of Antarctic desert soils on Earth. The Antarctic desert is a good benchmark for conditions on Mars, except that Mars is even drier and colder.

This brief discussion of Martian and lunar soils has only one purpose: to establish the fact that soil, as currently defined in Soil Taxonomy, does occur on other worlds. We can be reasonably sure, however, that the formation and evolution of soils on Earth bears little resemblance to the same processes on other celestial bodies. Although soils exist on the Moon, on Mars and on Venus, the conditions under which they formed are drastically different from Earth. Therefore, this book concerns itself exclusively with Earth, leaving the soils of other planetary bodies to another time and to other authors.

CHAPTER 2. THE EARTH'S SURFACE: A HISTORY

Beginning about 1890, scientists began gathering data from around the world to determine the chemical makeup of the Earth's surface. F.W. Clarke of the United States Geological Survey was an early leader of this effort. Clarke, a chemist, teamed up with H.S. Washington, a geologist, and together they evaluated more than 5,000 rock samples from different parts of the world. From these data they calculated an average elemental composition of the Earth's crust. Other scientists conducted similar analyses over the years, but the overall results have been about the same; as shown in the table below, just eight chemical elements make up nearly 99% of the Earth's crust by weight.

Note that oxygen accounts for nearly half the weight of the Earth's surface, and the element is even more impressive on a volume basis. The oxygen ion is so large that, on a volume basis, it accounts for more than 90% of rock structure. According to Parker (1984), the Earth's surface rocks are "for all practical purposes solid oxygen with a few impurities." A piece of granite or other rock can be viewed as an oxygen matrix or framework in which "impurities" such as calcium, potassium, and magnesium are scattered about. But simply listing the elements found near the surface of the Earth, the elements from which mineral soils are derived, is not sufficient for our purposes. In order to gain a more complete understanding of how rocks break down and turn into soils, we must know more. For example, how were the elements created, how did they end up on Earth in such quantities, and how did certain elements come to dominate the surface? We can gain some insight into these questions by briefly reviewing the history of our planet, with special emphasis on how the surface evolved over a period of 4.5 billion years.

Table 2.1. Elemental content of the Earth's crust

Data from Clarke (1924), Mason (1952) and Vinogradov (1956)

Element	Percent by weight
Oxygen (O)	46.8
Silicon (Si)	28.6
Aluminum (Al)	8.1
Iron (Fe)	4.8
Calcium (Ca)	3.3
Sodium (Na)	2.7
Potassium (K)	2.5
Magnesium (Mg)	2.0
Total	98.8

Carl Sagan wrote in *Cosmos* (1980) that, "If you wish to make an apple pie from scratch, you must first invent the universe." What is true of apple pies also is true of soil, so the best way to tell the story of soil is to begin with the event responsible for its existence and for the existence of us and everything around us, the Big Bang. No one is sure what, if anything, existed before the Big Bang. A general consensus among cosmologists is that nothing existed; but not only did nothing exist, there was no space in which anything could have existed, since space itself did not yet exist. Suddenly, out of this profound nothingness, the universe came into being. According to one pundit, "First there was nothing; then it all exploded." Even the most able scientists struggle to avoid sounding mystical when describing what happened, because the sudden creation of everything out of (supposedly) nothing is so counterintuitive, so contrary to everyday human experience.

No one can explain fully why and how the Big Bang happened, but nearly a century of research has provided considerable detail as to what took place immediately afterwards. Electrons, protons, and neutrons, the entities from which all matter would eventually form, were created during the first few seconds. Initially the expanding universe was too hot for them to come together; but after half a million years or so, things had cooled down enough for the first elements to form; mostly hydrogen, but also small amounts of helium and traces of lithium. These, of course, are the first three elements in the periodic table. Hydrogen has one proton, helium has two protons, and lithium has three, so they are elements one, two, and three respectively.

If not for the force of gravity, element creation would have stopped there. But, as Hazen (2012) expressed it, "Gravity is the great engine of cosmic clumping," and this simple process of clumping was to have a profound impact on element creation and the subsequent evolution of the universe. Individual hydrogen atoms are infinitesimally small, but not too long after the Big Bang, they were created in staggering quantities. Very soon, in cosmic terms, hydrogen clouds began to collapse as vast numbers of hydrogen atoms were pulled toward a common center, forming giant, rotating clouds with rapidly growing pressure and heat at their centers. The core temperatures of these stars-in-the-making soon reached millions of degrees, with pressures exceeding millions of Earth atmospheres. Under such extreme conditions, hydrogen nuclei began slamming together, and as the number of collisions grew, hydrogen atoms began to fuse, creating helium atoms. Each helium atom created was about one percent less massive than the combined mass of the four hydrogen atoms that were required to make it. The missing mass had been converted to energy, as predicted by Einstein's formula, $E = mc^2$. As fusion continued and as matter was increasingly converted into energy, the first stars began to radiate heat and light into the frigid space around them.

Most of us learned at an early age that atoms are made up of protons, neutrons, and electrons. Although all parts of the atom are important, when it comes to defining elements, only protons count. All atoms with seven protons, for example, are nitrogen. Similarly, any atom with six protons is by definition a carbon atom. New elements are created in stars by slamming lighter atoms together with such force that they fuse together, packing more protons into the nucleus and creating newer, heavier elements. Not surprisingly, this process is referred to as nuclear fusion. Element creation in stars is a sequential process, with smaller atoms slamming together to make helium, then carbon, and so on until all of the first 26 elements in the periodic table have been created (Burbidge et al. 1957).

In every case, with every element from helium (element number 2) up to iron (element number 26), the fusion process liberates energy, but no energy can be released after iron, which has been dubbed the "ultimate nuclear ash" (Hazen 2012). The appearance of iron indicates that a star is nearing the end of its life cycle. Shortly after forming an iron core, large stars destroy themselves in cataclysmic explosions. When fusion comes to a halt, energy production stops and heat is no longer generated in the star's core. At this point gravity takes over, trying to pull every atom of the star to the center, with no opposing force to stop it. Soon the star collapses in upon itself with such incredible force and speed that it is ripped apart, blasting most of its immense mass out into space. No one knows when the first such explosion occurred, but soon after the universe was created, stars were forming, living out their lives and then exploding violently throughout the universe. Each of these explosions hurled the remnants of dying stars, including the new elements they had created, out into space, where they intermixed with the vast clouds of hydrogen that had thus far evaded the star-forming effects of gravity.

How is the creation of elements in stars related to the nature of modern soils? As mentioned previously, just eight elements—oxygen, silicon, aluminum, iron, calcium, magnesium, potassium, and sodium—make up nearly 99 percent of the rocks at the Earth's surface, and these are the rocks with which soil formation begins. Table 2.2 shows the number of protons in the nucleus of each of these elements.

Table 2.2. Elements dominating surface rocks and the number of protons in their atoms

Element	Number of protons
Oxygen	8
Sodium	11
Magnesium	12
Aluminum	13
Silicon	14
Potassium	19
Calcium	20
Iron	26

None of these eight elements has more than 26 protons, meaning that all of them are readily formed during the life cycles of large stars. But something is missing here; there are more than 90 naturally occurring elements on Earth and we have accounted for the creation of only a few. What about all of the others, the elements beyond iron? Very simply, these larger elements are forged in the extreme heat and violence that ensue during the short time that a giant star is exploding. As a result, heavier elements such as gold, silver, lead, and uranium are much less common on Earth than are the 26 elements formed so abundantly by stars during the normal, more controlled stages of their life cycles.

The Solar System

The story of our solar system began more than 4.5 billion years ago with a great nebula of hydrogen gas, ice, and dust stretching across a vast expanse of space about halfway out from the center of the Milky Way. Suddenly something happened; perhaps a nearby star exploded, sending shock waves out into space and destabilizing the nebula. In any event, gravity began to pull the hydrogen gas, ice, and dust particles toward the center of the nebula. As the vast cloud of matter collapsed, it began to rotate, then to spin faster and faster, similar to what happens when an ice skater folds her arms close to her body. Eventually the spinning nebula assumed the shape of a flattened disc with a bump or bulge at the center, much like popular renderings of flying saucers. That central bump or bulge was the beginning of our sun. As it grew in size, the gravitational field of the developing sun grew stronger and stronger, eventually absorbing more than 99 percent of the material that had made up the original nebula. As the sun gained mass, the temperature and pressure at its center began to increase, in time becoming intense enough to begin fusing hydrogen atoms together, igniting the solar furnace.

Even after the sun had gathered in more than 99 percent of everything that existed in the pre-solar nebula, an extensive cloud of ice, dust, and gas still remained. This orbiting cloud of matter, from which the planets would form, extended out to a distance of five to six billion kilometers (km). We know this because Pluto, one of the planet-like bodies formed from this nebula, is six billion kilometers from the sun.

The fact that the pre-solar nebula was spread over so much space is one reason that our sun has so many planets. Interestingly, the planets are not traveling through space at the same speed; those nearest the sun orbit much faster than those farther out. Mercury, closest in, is traversing space at a speed of about 170,000 kilometers per hour, more than 1.5 times as fast as Earth, while Pluto, billions of kilometers more distant, travels at the comparatively slow speed of about 17,000 kilometers per hour, only one tenth as fast. The other planets follow the same pattern, as shown below.

Readers wishing to do so can easily convert kilometers to miles; simply multiply the number of kilometers by 0.62.

Table 2.3. Orbital velocities of the planets (rounded to nearest 1,000 km/hour)

Planet	Orbital velocity (km/hour)
Mercury	172,000
Venus	126,000
Earth	107,000
Mars	87,000
Jupiter	47,000
Saturn	35,000
Uranus	25,000
Neptune	20,000
Pluto	17,000

With respect to orbital speed, what is true of the planets would have been true of the clouds of matter from which they formed. Before the planets were created, parts of the clouds nearest the nascent sun would have been orbiting much faster than material a billion or more kilometers farther out. As particles and gas continued to travel around the sun, gravity and random collisions would have started the process of cosmic clumping. Whenever two particles collided, the larger would nearly always win and end up absorbing the smaller. This game of winner take all would continue until nearly all the material in a given orbital range had been swept up and consolidated into one massive body. For example, Earth became the big winner in its area of gravitational influence, but it was unable to attract material beyond a certain range. That mass was lost to Earth's two nearest competitors: Venus, closer to the sun, and Mars, farther out.

Earth had attained its present mass by approximately 4.5 billion years ago, but conditions then gave no hint of what it was to become. The earliest phase of Earth's history is referred to as the Hadean or Hellish Eon, and hellish it was. Earlier in this chapter we discussed the way in which gravity creates immense heat and pressure in the interior of stars as it tries to compress more and more matter into a smaller and smaller volume. A similar process takes place during the formation of planets, but the heat and pressure are not nearly so great. But a lot of heat is created in the interior of planets, so much that the surface of the early Earth was a roiling cauldron of molten rock, made even more hot and chaotic by the decay of radioactive elements and by the impact of asteroids and comets striking the surface almost continuously. The recently formed moon, then at a distance of only 15,000 miles from Earth, added to this hellish aspect by generating tides of hot lava a mile or more in height, tides which were constantly circling the globe unimpeded by any landmass. Adding to the chaos and disorder, the Earth at that time was spinning at a dizzying rate, with the sun rising every five hours.

But eventually this hell on Earth came to an end. Prodigious amounts of heat were continuously radiated out into space, so surface cooling was inevitable, and sometime before Earth was 100 million years old, the surface had "frozen" into a thin black crust consisting of basalt. It might seem odd to think of material solidifying at temperatures measured in thousands of degrees Fahrenheit as freezing, but in concept it is little different from ice forming on the top of a lake. This short-lived phase of our planet's history is sometimes referred to as "Black Earth." As our planet continued to evolve, the color of its surface would change a number of times; the Black Earth would become a Blue Earth, a Gray Earth, a Red Earth, and finally, a Green Earth. This rough color sequence reflects the evolution of the planet's surface (and of its minerals, soils, and life forms) over several billion years.

The basalt making up Earth's first solid surface was buoyant enough to rise to the surface but too heavy relative to the underlying material to develop any dramatic topography. Scattered volcanic cones would have stood out as isolated peaks, but nearly everything in between would have been a monotonous terrain of flat lands and low hills; there would have been no great mountain ranges, no Grand Canyons. The Black Earth was destined to last for only a short time before being transformed into a Blue Earth, a watery world with a globe-encircling ocean a kilometer deep or more. By the time Earth was 200 million years old, the original basalt surface had become the floor of a vast ocean.

The water that covered Earth was literally spewed out from the planet's interior. For millions of years, mega volcanoes had pumped prodigious amounts of water vapor into the atmosphere, and as the surface magma was cooling, so was the super-saturated atmosphere above. We do not know exactly when the downpours began, but sometime before its 200 millionth birthday, Earth began to be drenched by torrential rains, accompanied by hurricane-force winds and tumultuous thunder as giant bolts of lightning lit up the sky. The rain fell and continued to fall, perhaps for centuries, perhaps for hundreds of centuries. Low-lying basins started to flood, then grew into giant lakes and then into oceans, until the entire surface of Earth was covered with a layer of water a kilometer deep or more.

Volcanoes continued to erupt beneath the oceans, spewing basalt into the surrounding water and creating chains of volcanic peaks and islands. This oceanic world dotted with volcanic islands had an air of permanence, but things already were happening beneath the water and beneath the basalt floor on which it rested, things that would transform the surface yet again. It took millions of years, but eventually a new kind of rock started to rise up out of the water. This new rock was a dull gray, with distinct grains of much lighter and much darker material. Granite had made its appearance, and once this new rock got started, there was no stopping it.

The Origin of Granite

Covering the Earth's surface with a mile or more of water was much like putting a lid on a slowly boiling pot. The underlying rocks were reheated from below, and in many areas began to melt, but not all at once and not completely. The newly created magma cooled slowly, forming rock that was chemically and physically different from basalt. It had larger crystals, was lighter in color, was less dense, and had become relatively enriched with some chemical elements and relatively impoverished in others; it had become granite, a much lighter rock which would easily float on top of the viscous material beneath. The table below shows some representative values for the elemental makeup of basalt and granite.

Table 2.4. Elemental composition (percent by weight) of some granite and basalt samples

Calculated from the data of Clarke (1924) and Harrison (1933)

Element	Basalt	Granite	Percent gain or loss
Iron	6.1	2.3	-62%
Calcium	8.8	1.3	-85%
Magnesium	4.7	0.4	-91%
Potassium	1.5	3.4	+127%
Sodium	1.7	2.7	+59%
Silicon	23.6	32.9	+39%

The trends are quite easy to see. Compared with basalt, granite is much poorer in iron, calcium, and magnesium and much richer in silicon, potassium, and sodium. It should not be surprising that granite, in comparison with basalt, contains relatively more of the lighter sodium and potassium and relatively less of the heavier iron, magnesium, and calcium. It also should not surprise us that granite overall is less dense than basalt. That is why granite made its way inexorably to the surface to form continental landmasses floating on the denser material underneath like giant icebergs. As basalt and the rocks beneath it continued to partially melt and then refreeze, layer after layer of buoyant granite piled up, then began to rise high above the oceans. The Rocky Mountains, with peaks higher than 4,200 meters (14,000 feet) and granitic roots extending to depths of nearly 80 kilometers (50 miles), are a striking example of this process and a visible testament to the fact that granite floats. The difference in buoyancy between granite and the underlying mantle is the key to continent formation.

Before moving on, some clarification is in order. A main point of this chapter is that granite, being less dense than basalt, worked its way to the surface to form the continents. This suggests that large areas of exposed basalt would be rare or nonexistent on land surfaces. However, if you live in an area such as America's Pacific Northwest or the Deccan Plateau of India, you know this is not true. Beautiful surface exposures of basalt can be seen at a number of locations throughout the world. Basalt can form whenever lava reaches the surface and then cools rapidly, whether under the ocean or on land. Such basalt flows on land, commonly referred to as flood basalts, are largely a surface phenomenon; lava flows out of a volcano, spreads across the land surface much like pancake batter, and then hardens rapidly to form a veneer of basalt. Areas covered by prehistoric flood basalts tend to form rich soils of great agricultural potential, but they are regional phenomena and do not negate the general concept of a basaltic ocean floor and lighter, floating continents rooted in granite.

Granite continued to form, rise to the surface, and add steadily to the continents, and by the time Earth was about 1.5 billion years old (three billion years ago) the surface consisted of about two thirds ocean and one third land. Because of plate tectonics, these landmasses kept moving about, occasionally coming together to form supercontinents and then breaking up again as the underlying plates drifted apart. The atmosphere at that time is believed to have consisted of water vapor, carbon dioxide, and nitrogen, along with lesser amounts of other gases such as hydrogen, methane, and ammonia. Carbon dioxide levels were higher than today, so the rain would have been more acidic, since atmospheric carbon dioxide readily dissolves in rainwater to produce a dilute solution of carbonic acid.

Under these conditions, feldspar, a major component of granite, would have weathered to form clay, breaking down the rock in the process.

There would have been soils, but nothing like the soils of today. There was no free oxygen in the atmosphere to oxidize iron, so the bright reds, browns, and yellows so striking in modern soils would have been absent; everything would have been uniformly gray. In addition, these early soils, or proto-soils, would have been shallow and lifeless: no plants, no roots, no fungi — perhaps a few microbes, but none of the living things that enable modern soils to play such an important role in the economy of nature. These shallow, dull, lifeless precursors were just a hint of the soils to come, but modern soils were not destined to make their appearance for at least another two billion years. Several important things had to happen first.

The Great Oxidation Event

One of them was already happening. While the continents were slowly forming, life had been busily evolving in the oceans. Scientists do not agree as to the exact timing, but most think that by the time Earth was a billion years old, life had become firmly established in the seas. For a very long time, all living things were single celled and were confined to the oceans. Eventually, after much trial and error, some of these organisms mastered a very useful skill. They learned how to harness the power of the sun and use it to manufacture a molecule we now call glucose from water and carbon dioxide:

$$6\ CO_2 + 6\ H_2O + \text{energy from the sun} \Rightarrow C_6 H_{12} O_6 \text{ (glucose)} + 6\ O_2.$$

Organisms able to carry out this marvelous process flourished and soon were producing glucose on a gargantuan scale and releasing vast amounts of oxygen "waste" into the surrounding water. For a long time, much of the oxygen released was used up in oxidizing the huge amounts of iron that had been dissolved in the world's oceans. This process created most of the recoverable iron ore that later would make the Industrial Revolution possible. But iron oxidation was not the only thing delaying the oxidation of Earth's early atmosphere. For several hundred million years, microbes living in the ocean used up immense amounts of oxygen in converting ammonium (NH_4) to nitrate (NO_3), which other microbes then converted to nitrous oxide (N_2O) and nitrogen gas (N_2). But eventually oxygen gas began to escape from the oceans and accumulate in the atmosphere.

This had a profound effect on the world's rocks and soils. Very quickly, at least in geologic terms, rock surfaces and soil surfaces turned red. This change came about as atmospheric oxygen reacted rapidly with iron in surface minerals to create iron oxide (Fe_2O_3). The advent of free oxygen to the atmosphere was destabilizing, causing surface rocks to weather more rapidly and to form soils that were a bright red instead of gray. When iron atoms in a rock react with oxygen to become iron oxide, the rock starts to fall apart, so it is no surprise that iron-bearing rocks are among the first to show the effects of surface weathering. This oxygen-induced color change was spectacular, but the more important change, the one with the most far-reaching implications, was simply the accumulation of free oxygen in the atmosphere. This "Great Oxidation Event," as it has been called, in addition to increasing the rate of rock weathering, would make the existence of land animals possible; it was a critical milestone in the evolution of Earth's surface.

With the advent of free oxygen, the world's soils began to take on a distinctly modern look, but some vital ingredients still were missing, such as plants, organic matter, and roots. It would be a long time, another two billion years or more, before life would venture out of the ocean and begin turning the red Earth to green. Fossilized spores tell us that primitive plants had colonized the land surface by around 470 million years ago. These unimpressive pioneers were only a few inches high and had no leaves, only greenish stems and branches to capture the sun's energy. Lacking roots or stems to transport water, these pioneers would have been restricted to low, wet areas; but by 400 million years ago, more complex species had arisen and were spreading out across the once-barren land. But they had a vascular system and rudimentary roots that anchored them in the ground, enabling them to absorb water and nutrients.

No one is certain when the first tiny leaves evolved, but by vastly increasing the efficiency of photosynthesis, they largely defined the nature of plant competition from then on. In order to be successful, a plant species had to win the struggle for sunlight, and plants that were able to grow the tallest and produce the most leaf area had a competitive advantage. This mad scramble for sunlight and energy soon led to the evolution of tall, woody stems, branching limbs, and larger, greener leaves. Another epic competition was being waged underground, spurring the development of more extensive and more efficient root systems. By 360 million years ago, magnificent forests had appeared and much of the land surface was clothed in greenery. More extensive root systems, with their bumptious, probing populations of root tips increased the weathering rates of rocks by several orders of magnitude. Soils grew deeper and richer in clay, and soon they began to develop the distinctive horizons that characterize so many soils today.

Most importantly, the soils came alive. As roots died and decayed and as dead leaves, stems, and eventually dead trees were deposited on the surface, soils became enriched in organic matter and provided the carbon, nitrogen, sulfur, and other elements needed for life to flourish. And flourish it did. The soil, soon populated with untold species of microbes, fungi, insects, worms, and even small reptiles and mammals, became a biological treasure house. This dynamic, living boundary where the lithosphere, atmosphere, and biosphere meet began to play an increasingly important role in world ecology. Modern soils, teeming with life, stored and released nutrients and water for plant growth; exchanged elements such as oxygen, carbon, nitrogen, and sulfur with the atmosphere; and served as a source of nutrients for the complex ecosystems of rivers, lakes, and oceans.

When one considers the sequence of events that led from the Big Bang to the formation of Earth and then the stepwise evolution of the Earth's surface, much of it seems predestined or preordained. A presumptuous notion indeed, but geologist Robert M. Hazen, in his book *The Story of Earth* (2012), argues that there is indeed such a thing as "elemental inevitability." Professor Hazen writes,

> The production of large numbers of protons and electrons, and of corresponding amounts of hydrogen and helium, was hardwired into our universe from the instant of the Big Bang. The formation of stars was an inescapable consequence of the production of huge amounts of hydrogen and helium. The synthesis of all the other elements by nuclear fusion reactions and by supernovas was equally preordained by the formation of hydrogen-rich stars.

Hazen further argues that the accretion of material hurled into space by exploding stars and the conversion of this material into yet another generation of stars, some orbited by planets such as Earth, also were pre-ordained. The fact that science works is a strong argument that the universe is not all randomness and chaos; that instead there is a comforting level of determinism that makes many events predictable and thus understandable. Once you know certain things, much of what happens in cosmology and geology, and in other sciences as well, is predictable, or at least partly predictable. The qualifier "partly" must be used because superimposed upon this core of predictability is just enough randomness or chaos to make things interesting. For example, this chapter highlighted the stepwise evolution of our planet's surface, noting how the molten surface cooled to form a thin black crust, which soon was covered by a worldwide ocean. We then described how the gray granitic continents rose up out of this ocean and remained gray for a long time, until ocean dwelling algae sent a flux of oxygen into the atmosphere and turned them red. We then described how, later still, the plant kingdom ventured out of the ocean, colonized this red land and clothed it in green. In retrospect, knowing what we now know about the chemical elements as well as gravity and the other laws of physics, much of this progression was predictable.

This chapter also made much of the creation of granite from the partial melting and recrystallization of denser rocks and how this formed the continents. But here is a question that is worth pondering: If the continents are made up mostly of granite, why are there so few places in the world where we can see granite at the Earth's surface? It has been estimated that igneous rocks such as granite are found at the surface on only about 20 percent of the Earth's land area. The rest is buried by sedimentary rocks such as shale, sandstone, and limestone, or by water and wind-deposited materials such as river alluvium, glacial till, and loess. Much of America's northeast and upper Midwest, for example, is covered with many feet of glacial till, while large parts of states such as Nebraska, Kansas, Missouri, Iowa, Illinois, and Tennessee are buried in loess. Large portions of eastern Washington also are loess covered, while much of western Nebraska consists of ancient sand dunes now anchored by grass. Deep layers of water-deposited sediment make up the Southeastern Coastal Plain, the Mississippi River floodplains, and the Central Valley of California.

So much of the Earth's surface now consists of sedimentary rocks, alluvium, glacial till, loess, and wind-blown sand because, once Mother Nature had created the continents, she was not content to leave well enough alone. Instead, she proceeded in a number of nondeterministic, unpredictable ways to use ice, water, and wind to reshape the land, grinding rocks up into small particles, moving the debris far and near, and then depositing it again on the surface. Much of the North American land surface was reshaped only a few thousand years ago by the direct action of glaciers and by the vast amounts of turbulent water and frigid hurricane-force winds that accompanied them. This reshaping was a gift of almost unimaginable value to America, leaving us with huge expanses of prime farmland that are unmatched anywhere in the world. For modern Americans, recent glacial activity and its associated effects were very good things indeed. However, for the purposes of this book, it does present us with a minor bookkeeping problem.

Table 2.1, at the beginning of this chapter, shows the relative amounts of elements making up the Earth's crust, a layer many miles thick. But considering the extent to which the Earth's near surface has been disturbed, these numbers simply do not work for us. Soil formation, by definition, takes place at the very surface; so instead of knowing about the entire crust, we must know which chemical elements are predominant in the first 50 feet (15 meters) or so. Although the crust as a whole is mostly igneous in origin, sedimentary rocks such as sandstone, shale, and limestone are highly concentrated in the upper part, covering nearly 80 percent of the Earth's land surface. According to one estimate, the near surface (the part of the crust where soils form) consists of about 20 percent igneous rock, 42 percent shale, 20 percent sandstone, and 18 percent limestone (Ronov and Yaroshevsky 1972). As shown in the table below, the chemical makeup of these four rock types varies considerably.

Table 2.5. Major elements in different rock types (percent by weight)

Data from Parker (1967)

Element	Igneous	Shale	Sandstone	Limestone
Silicon	27.8	27.3	36.8	2.4
Aluminum	8.1	8.2	2.5	0.4
Iron	5.1	4.7	0.8	0.4
Calcium	3.6	2.2	3.9	30.2
Sodium	2.8	1.0	0.3	0.1
Potassium	2.6	2.7	1.1	0.3
Magnesium	2.1	1.5	0.7	4.7

Using the data from Table 2.5 and taking into account the degree to which the Earth's surface has been disturbed, a weighted average content was calculated for each major element in surface rocks (Table 2.6).

The chemical aspects of rock weathering and soil formation largely revolve around the properties and ultimate fates of these eight elements. Oxygen makes up more than 50% of surface rocks by weight and well over 90% by volume. As rocks weather, some elements are easily removed from the oxygen matrix and carried away, while others mostly stay in place to become the building blocks of soil. Once we understand the way in which each element reacts to the weathering process, we can begin to see how rocks become mineral soil, the main topic of the next chapter. It will become increasingly clear that rock weathering and soil formation, like so many of the geological processes described in this chapter, are comfortingly predictable.

Table 2.6. Weighted elemental content of near-surface rocks

Element	Percent by weight
Oxygen	52.4
Silicon	24.8
Calcium	7.9
Aluminum	5.6
Iron	3.2
Magnesium	2.0
Potassium	1.9
Sodium	1.0
Total	98.8

This chapter has emphasized the pervasive role that gravity plays in the creation of elements in stars. It is rather astounding that this single force, starting with just hydrogen (plus a little helium and lithium), was able to create all of the chemical elements that now make up our universe. This chapter also describes the way in which the Earth developed a layered structure as gravity pulled heavier elements like iron to the center while lighter elements such as silicon rose to the surface to form the floating continents.

Gravity is one of the four fundamental forces of nature; the other three are the strong nuclear force, the weak nuclear force and electromagnetism. All of these forces came into being during the first chaotic seconds after the Big Bang. The strong nuclear force and the weak nuclear force operate within the nuclei of atoms. Although they are absolutely essential for holding the world together, there is no need to concern yourself with them unless you are a nuclear physicist. But gravity and electromagnetism are different — everything that we see happen in the world around us, from falling rain to the formation of soil, can be attributed to the effects of just these two forces, sometimes acting alone, sometimes together and sometimes at cross purposes. Interestingly, the force of electromagnetism is about 10^{38} times stronger than that of gravity, but gravity is additive, so very large masses can generate very powerful gravitational fields that act over immense distances. For example, Pluto is six billion kilometers out in space, but it is still held in place by the gravitational pull of the sun.

The next chapter discusses the chemical makeup of rocks and the agents that transform them into mineral soil. Much of that discussion focuses on electrons and the chemical interactions between them, clearly in the realm of electromagnetism. Although this book is nominally about rocks, rock weathering and soil formation, there is a subtext; every process or phenomenon described also is about gravity and/or electromagnetism, because they make it possible for everything to happen. Science consists of trying to figure out how they do it

Chapter 3. Rocks and Rock Weathering

In describing rocks, we employ words such as solid or hard almost without thinking, because rocks are among the hardest things we encounter in the world around us. It is no accident that the Stone Age marks an important period in human history, a time when our ancestors became skilled at turning rocks into tools and weapons. But here is a surprising fact about rocks: Although they are indeed hard, they are far from solid, at least at the atomic level. The stony surface beneath your feet and the rocks that feel so heavy in your hands are mostly empty space. Rocks are not unique in this respect; trees, grass, animals—all of the "solid" objects we see around us—are the same. Everything is made up of atoms, and all atoms consist of tiny bits of matter occupying a large volume of empty space.

Consider an atom of oxygen (O), which accounts for more than half the mass of all rocks on Earth. Oxygen is element number eight, meaning that it has eight protons and eight neutrons (in most cases) in the nucleus and eight electrons orbiting at a distance. Electrons are extremely small; it takes more than 1,800 of them to equal the mass of just one proton. An appropriate spatial model for the oxygen atom might be 16 bowling balls glued together to form a nucleus (representing the eight protons and eight neutrons) with eight tiny glass beads (the electrons) orbiting at a distance of several miles. The atoms of other elements are arranged the same way, with tiny electrons orbiting at a vast distance from the nucleus. With so much space between the nucleus and its cloud of spinning electrons, it is not surprising that the nuclei of atoms rarely come into contact; but the electrons from different atoms are always bumping into each other. That is what much of chemistry is about. Chemical reactions take place when two or more atoms meet and combine by sharing or exchanging electrons. Atoms will join together if doing so makes them more stable, and that is where a concept sometimes called "magic numbers" comes into play.

Magic Numbers

A further discussion of oxygen will help to understand the concept of magic numbers. For the first two billion years or so of our planet's history, the atmosphere was devoid of free oxygen; but that is no longer true. Thanks to green plants, this vital element now makes up about 21 percent of the atmosphere. But despite millions of years of photosynthesis, more than 99 percent of all the oxygen on Earth still remains locked up within the structure of rocks. What is it about oxygen that causes it to combine so readily with elements such as silicon, aluminum, and iron to form rocks? The answer lies in the number of electrons; oxygen has eight, and that single fact accounts for much of its chemical behavior.

Atoms often are described as being like tiny solar systems, with electrons orbiting around the nucleus much as the planets revolve around our sun; but there is a big difference. If Earth were a thousand miles closer to the sun or a thousand miles farther away, it would not matter so much. Things might be a little different, but the solar system would function much as it does now, and no fundamental rules of nature would be violated. It is different with atoms; the electrons must obey certain rules, rules first discerned by Danish physicist Niels Bohr in 1912. Bohr, while still a student, figured out the basic structure of what is now referred to as the Bohr atom. The young scientist postulated that electrons do not orbit randomly around the nucleus; instead, they are restricted to certain paths or energy levels. Not only must electrons travel in predetermined energy levels, there are strict rules about how many electrons are allowed to orbit in each path or energy level. The first energy level out from the nucleus allows a maximum of two electrons, while the second and third levels allow a maximum of eight electrons each (2, 8, 8). If an atom has even more electrons, a fourth shell begins to fill, and this shell allows a maximum of 18 electrons, so we are up to 2, 8, 8, and 18.

The way electrons fit into shells enabled Russian chemist Dmitri Mendeleyev to place all of the known elements in a logical, sequential order in what came to be called the periodic table. When Mendeleyev introduced the periodic table in 1869, Niles Bohr had yet to be born, so nothing was known of the Bohr atom or of electron shells. Mendeleyev simply recognized that by assigning each element to a box, with the atomic number of the elements increasing from left to right, he was able to orga-nize the table into columns in which all of the elements within a column reacted chemically in roughly the same way.

The discussion to come will be made much clearer if readers have a copy of the periodic table available as they read. If you key a phrase such as "images periodic table" into a search engine, you soon will be presented with many examples that can be printed out for further reference. Internet sources are used throughout this book to illustrate important points or to enable readers to see good examples of things being discussed. Assuming that you have acquired a copy of the periodic table, let us continue. The column farthest to the right in the table (column 18) is where the so-called noble gases — helium, neon, argon, and so on — reside. All of the elements in this column are inert, meaning that they do not react chemically with any other element. Reading top to bottom, helium has a total of two elec-trons, neon has ten, argon has 18, krypton has 36, and so on. These are the "magic numbers." Helium, with the magic number of two electrons, has the first energy level completely filled and thus has no reason to exchange or share electrons with other atoms. If a helium atom encounters a carbon atom, for example, it will simply bounce off and continue on its way. Neon, directly below helium, has a total of ten electrons. With this magic number, neon has the first energy level filled with two electrons and the second filled with eight electrons (2, 8). Like helium, it will simply bounce off of any other atom it happens to encounter; in other words, it is inert or inactive. Continuing down the column, argon, with 18 electrons, has an electron configuration of 2, 8, 8 — so once again all of the shells are filled. Argon, like helium and neon, does not engage in chemical reactions with other elements.

Let us return to oxygen, which has a total of eight electrons, two in the first shell and six in the second (2, 6). To be stable, the oxygen atom really needs two additional electrons in that last shell to assume a (2, 8) configu-ration. Because of this, the oxygen ion is designated as O^{2-}. The 2- means that the oxygen atom, with eight electrons, is constantly on the lookout for two more in order to reach the stable magic number of ten, so that it will have two electrons in the first energy level and eight in the second energy level (2, 8). Now let us turn to the other seven elements that, in addition to oxygen, dominate the structure of rocks. Table 3.1 lists the elements, their number of electrons, the nearest magic number, and the number of electrons each element must acquire or lose to reach magic number status.

Table 3.1. Total electrons, nearest magic numbers, and excess or deficit of electrons in elements dominating surface rocks

Element	Total electrons	Nearest magic number	Excess or deficit of electrons
Oxygen	8	10	-2
Sodium	11	10	+1
Potassium	19	18	+1
Calcium	20	18	+2
Magnesium	12	10	+2
Aluminum	13	10	+3
Silicon	14	10	+4
Iron	26	18	+8

Oxygen (O^{2-}) is the only major element in surface rocks that needs to acquire electrons in order to become stable. Not surprisingly, oxygen has been referred to as Earth's "master electron acceptor." Since the other major elements in rocks need to donate electrons to reach stability, they all readily combine with oxygen. Silicon stands out because it, like oxygen, is present in such large amounts, accounting for nearly one fourth of all atoms in the Earth's crust. In addition, it has four electrons to donate. It is no surprise that oxygen atoms eagerly combine with silicon. As a result, resistant silicon-oxygen bonds are found in almost every kind of rock; quartz, or SiO_2, accounting for most of the trillions of sand grains found on our beaches, is a common example. While oxygen is Earth's "master electron acceptor," all of the other elements listed above are electron donors, eager to give their surplus electrons away to oxygen.

Iron deserves special mention. Unlike the other elements in rock, which need to give away one, two, three, or even four electrons, iron has a total of eight surplus electrons. It has 26 and would like to reach the nearest magic number of 18. No element in the entire periodic table will accept that many, so iron does the best it can. It gives away as many electrons as possible, usually to oxygen. In some cases, oxygen will accept two electrons, forming FeO, while in others it will accept a total of three, resulting in the formation of Fe_2O_3.

Oxygen and Ionic Bonds

When oxygen combines with the other elements in rocks, it does so by way of chemical bonds that are mostly ionic. An ionic bond is one in which an element mostly gives up one or more electrons, as opposed to merely sharing them on a somewhat equal basis. When one element essentially assumes control of an electron from another element, both of the elements become ions. In the case of silicon and oxygen, silicon becomes the Si^{4+} ion, while each of the oxygen atoms becomes an O^{2-} ion. By convention, ions with excess electrons, such as Si^{4+}, are called cations, while ions with a deficit of electrons, such as O, are called anions.

When elements such as silicon and aluminum combine with oxygen to form rocks, the structural arrangements are governed by Pauling's Rules. These five rules, laid out by Linus Pauling in 1929, are a very useful guide to understanding rock structure. Note: It is important to remember the distinction between rules and laws. Laws, such as the law of gravity, apply universally; there are no exceptions. Rules, although they apply most of the time, nearly always have exceptions. We will not use all five of Pauling's rules, but the first two are very applicable to this discussion. Rock structure consists basically of an oxygen matrix in which a small volume of cations such as Na^+ and Ca^{2+} are enclosed. Based on simple geometry, larger ions must surround themselves with more oxygen ions than smaller ones. But how many? Pauling's first rule, the radius ratio rule, enables us to answer that question. Table 3.2 shows the number of oxygen ions needed to balance each of the major cations in rock. Note that the larger the cation, the more oxygen ions surrounding it. The math is quite straightforward and based on simple geometry. Readers interested in Pauling's reasoning and calculations are referred to his 1929 paper or to his 1960 book.

Values for the radius of each ion in Table 3.2 are given in angstroms (Å). Named in honor of Swedish physicist Anders Jonas Angstrom, the angstrom is not officially part of the International System of Units. However, it is a widely used unit of measurement for atoms, molecules and other things in nature that are extremely small. The average diameter of an atom ranges from approximately 0.5 Å for hydrogen (the smallest element) to 3.8 Å for uranium (the largest natural element). The size relationships among different ions in Table 3.2 are easy to see and understand when expressed in angstroms, since the units are small numbers or fractions of small numbers. To provide some perspective on scale, it takes about 254 million angstroms to equal one inch.

Table 3.2. Ionic radii and the number of O^{2-} ions surrounding each major ion in rocks

Cation	Ionic radius (angstroms)	Number of O^{2-} ions
Si^{4+}	0.42	4
Al^{3+}	0.51	6
Mg^{2+}	0.66	6
Fe^{2+}	0.74	6
Na^+	0.97	8
Ca^{2+}	0.99	8
K^+	1.33	12

Once the number of enclosing oxygen ions is known, Pauling's second rule, the electrostatic valence rule, can be used to estimate the relative strength with which each element is held in the rock structure. This provides a rough indication of how susceptible each element is to being removed from the rock structure and carried away to the sea. For example, is magnesium held more tightly than sodium? Is silicon held more tightly than iron? The electrostatic valence rule can help answer such questions. It is based on a simple calculation: Divide the valence of the central ion by the number of O^{2-} ions enclosing it. For our purposes, valence is simply the number of electrons an ion wants to give away. The valence or charge of Si^{4+} is 4, for example, and it is enclosed by four O^{2-} ions in the rock structure, so the electrostatic bonding strength, as it is called, equals 1.00 (4/4). Similarly, the valence of Fe^{2+} is 2 and it is enclosed by six O^{2-} ions, so its relative bonding strength is 2/6 = 0.33. This suggests that Si in rocks is more resistant than Fe to being removed by weathering. Huggins and Sun (1946) calculated the approximate energies of formation for cation-oxygen bonds from a reference state of gaseous ions. The energy of formation becomes larger as the cation-oxygen bond became stronger. Values ranged from over 3,100 kilocalories per mole for the silicon-oxygen bond to 299 kilocalories per mole for the potassium-oxygen bond. Table 3.3 compares the relative values for bonding strength based on Pauling's simple rule with the calculated values of Huggins and Sun, with the Si-O bond arbitrarily set at 1.00 in both cases. The model predictions and the calculated values are in close agreement.

Table 3.3. Relative strength of bonds between selected cations and oxygen

Comparison of Pauling's rule with calculations by Huggins and Sun (1946)

Cation	Pauling's rule	Huggins and Sun calculations
Si^{4+}	1.00	1.00
Al^{3+}	0.50	0.58
Mg^{2+}	0.33	0.30
Fe^{2+}	0.33	0.29
Ca^{2+}	0.25	0.27
Na^+	0.12	0.10
K^+	0.08	0.10

The values in Table 3.3 are rough indicators of susceptibility to weathering, with large numbers, such as 1.00 in the case of silicon, indicating high stability and greater resistance. A smaller number, as in the case of sodium, indicates that the element is more easily removed from the rock structure. There are some definite trends. It appears that silicon and aluminum are held tightly in the oxygen matrix and thus would be highly resistant to weathering. They are followed by the more loosely held magnesium, iron and calcium, all of which appear to have about the same susceptibility. Sodium and potassium, at the bottom of the list, have very low values, indicating that they would be highly susceptible to weathering and removal from the rock structure.

But the numbers in Table 3.3 do not tell the whole story, since they apply only to the initial release of elements from rock. They say nothing about what happens after a sodium ion, a calcium ion, or any other ion is released into the near-surface environment. Our everyday experience suggests that sodium ions are easily leached from rocks and carried away to the sea; otherwise, the oceans would not be so salty; and there also is little doubt that large amounts of silicon and aluminum remain on the landscape to take part in soil formation. But we need more specifics. It is important to understand what happens to all of the elements after they are weathered from rock and what role each of them plays, if any, in the genesis of soil.

Rivers and Elements

We can gain some insight into this by examining the relative amounts of chemical elements dissolved in rivers. Geologists learned more than a century ago that rivers carry a large load of dissolved elements leached from the rocks making up a drainage area. Rain dissolves rock crystals much as water poured into a glass dissolves salt or sugar. Of course, the process is much slower for rock. Clarke (1924) estimated that for the United States, excluding the Great Basin, the land surface is dissolving away at the rate of about one foot or 25 centimeters every 25,000 years. This might seem extremely slow, but the Earth's surface is very old, so the rock-weathering agents have been at work for eons, time enough for an immense volume of minerals to have been weathered from rocks and carried away to the sea. The local effects of this global process can be quite impressive. The Niagara River, for example, carries about 3,600 tons of dissolved ions such as Na^+ and Ca^{2+} over its scenic falls every hour of every day. About 30 percent of the material carried by the Mississippi River at any given time consists of ions in solution and the average value for all the major rivers on earth is about 50 percent (Pringle 1985). The idea that rocks dissolve and continents weather away has been around for a surprisingly long time. Consider the following passage from King Henry IV, which historians believe Shakespeare wrote no later than 1597.

> And see the revolution of the times
> Make mountains level, and the continent,
> Weary of solid firmness, melt itself
> Into the sea.

The fact that the minerals dissolved in rivers reflect the chemical makeup of rocks in the drainage area can be readily demonstrated. Table 3.4 compares the dissolved loads of two rivers, the Chattahoochee in Georgia and the Shenandoah in West Virginia. The Chattahoochee drains an area of mostly igneous and metamorphic rocks such as granite, gneiss, and schist, while the Shenandoah drains an area underlain mostly by limestone. The dissolved loads of the two rivers reflect these differences in rock type. Note that the Chattahoochee has more than three times as much sodium, potassium, and silica (as H_4SiO_4) in solution as does the Shenandoah, while the Shenandoah contains more than twice as much dissolved carbonate (CO_3), calcium, and magnesium as the Chattahoochee.

Table 3.4. Dissolved loads (percent by weight) in the Chattahoochee River, Georgia, and Shenandoah River, West Virginia

Data from Clarke (1924)

Ion or compound	Chattahoochee	Shenandoah
CO_3^{2-}	21.32	47.22
Ca^{2+}	9.06	22.85
Mg^{2+}	1.51	5.86
Na^+	12.08	3.86
K^+	3.40	1.00
SiO_2	37.73	10.71
Al_2O_3	----	----
Fe_2O_3	1.13	0.07

Clarke (1924) compiled chemical data from rivers throughout much of the world and calculated the average amount of dissolved elements. Although some parts of the world were not well represented, Clarke's analysis included, among others, data from the Amazon, the Nile and the Mississippi. Considering the large and geologically diverse drainage areas of these three rivers, Clarke's data probably are representative of river chemistry worldwide. Using the data compiled by Clarke, the relative mobility of each major element in surface rocks was estimated using the approach pioneered by Polynov (1937), whose reasoning went something like this: Assume that, upon analyzing the rocks of an area, you find that a certain element occurs in only small quantities, but upon sampling the river(s) draining the area, you find that large amounts of the element are dissolved in the water. This would suggest that the element is very mobile, that it is easily leached from rocks and carried away. Now let us assume that a second element is very abundant in the rocks underlying an area but is almost nonexistent in the rivers that drain the area. This would suggest that this second element is not very mobile, that only small amounts are leached out of rocks and washed away. One also could infer that the first element is more mobile than the second. This rationale was used to estimate the relative mobility of all seven elements that combine with oxygen to form rocks.

Table 3.5 shows the average elemental content of the surface rocks of the world (from Table 2.4) and the average mineral content of the world's rivers (Clarke 1924). Note that sodium makes up only one percent of surface rocks, but it accounts for 5.8 percent of the dissolved load carried by rivers, suggesting that sodium is easily removed from rocks and carried away to the sea. The last column in Table 3.4 shows the average content of each element in river water divided by the average percentage of that same element in surface rocks. For example, the calculation for sodium was 5.8/1.0 = 5.80, indicating that the concentration of sodium is nearly six times greater in river water than in rocks. In contrast, the same calculation for silicon was 5.5/24.8 = 0.22, indicating that silicon is about five times more concentrated in rocks than in river water. In a relative sense, the mobility of sodium is more than 25 times that of silicon (5.8/0.22 = 26.4). This implies that sodium is much more susceptible to being leached from the land than silicon, a fact attested to by the very large concentration of sodium and the very small concentration of silicon in seawater.

Table 3.5. Estimated elemental content (percent by weight) of surface rocks and rivers of the world

Element	Surface rocks	Rivers	Rivers/Surface rocks
Sodium	1.0	5.8	5.80
Calcium	7.9	20.4	2.58
Magnesium	2.0	3.4	1.70
Potassium	1.9	2.1	1.10
Aluminum	5.6	1.5	0.27
Silicon	24.8	5.5	0.22
Iron	3.2	0.1	0.03

A relative mobility index for each of the rock-forming elements then was calculated, with sodium, the most mobile, set arbitrarily at 100. The numbers from the last column in Table 3.5 were used for these calculations. For example, the mobility of calcium relative to sodium was calculated by dividing 2.58 (the average amount of Ca in river water divided by the average amount in rocks) by 5.80 (the similar calculated value for Na). The answer, 0.45, indicates that, in a relative sense, calcium is less than one half as mobile as sodium. Similar calculations were done for the other rock-forming elements to derive a relative mobility index for each (Table 3.6).

Table 3.6. Relative mobility of the rock-forming elements

Element	Relative mobility
Sodium	100
Calcium	45
Magnesium	30
Potassium	20
Silicon	‹5
Aluminum	‹5
Iron	1

Considering the limitations of the data and the fact that these are global estimates, one should not make too much of the specific values in Tables 3.5 or 3.6, especially the small values for aluminum, silicon, and iron. As Keller (1957) cautioned, "The earth is a tremendously huge chemical laboratory and many parts of the apparatus are leaky." Later on, evidence is presented to show that the relative mobility of these three elements is, as shown in Table 3.6, Si›Al›Fe. That is, silicon is more mobile than aluminum, which in turn is more mobile than iron. To verify the overall results, a second analysis was conducted using a different, more recent set of data for river chemistry (Livingstone 1963), and not surprisingly, the numbers were slightly different. However, the direction and magnitude of the mobility rankings remained the same. But the true test of any analysis or any ranking is how well it explains or predicts events in the real world. As we continue, it will become increasingly clear that the relationships shown in Table 3.6 work and that they provide a simple but useful framework for understanding what happens to individual elements after rocks are destabilized and broken down by weathering agents.

Because the measurements were made nearly a century ago, some might question the use of Clarke's data on river chemistry and on the elemental composition of rocks. Chemical analyses were much more laborious in those days, but the laboratory methods were quite sound and the data have stood the test of time. In addition, using these data gave us the opportunity to honor one of America's premier scientists. F.W. Clarke has been called the father of geochemistry. He served as chief chemist of the United States Geological Survey for 42 years and was founder and first president of the American Chemical Society. His stature as a scientist is evidenced by the fact that almost a century after his death, the Geochemical Society still presents the annual F.W. Clarke Award to an early-career scientist who has made an outstanding contribution to the field.

Weathering Agents

The three main agents of chemical rock weathering are water, carbon dioxide, and oxygen. Of the three, water is the most important. If no carbon dioxide or oxygen were present, some chemical weathering still would take place from the action of water alone. But in the absence of free water, chemical weathering essentially comes to a halt because, in addition to being an active weathering agent, water also acts as a carrier for the other two. Carbon dioxide and oxygen do not function as weathering agents unless dissolved in water, which highlights how important it is that rain or water from melted snow moves down into the ground instead of running off the surface. The movement of water into the ground is enhanced if the surface rocks have been ground up or pulverized. Fortunately, this has happened on much of the world's land surface. For example, vast areas of what is now the United States are covered with deep deposits of glacial till, loess, or wind-blown sand, mainly as the result of glaciers and the turbulent water and hurricane-force winds that accompanied them.

Because of such widespread surface disturbance, much soil formation takes place, not in solid rock, but in what geologists refer to as regolith. The term regolith, first used by American geologist G.P. Merrill more than a century ago, comes from the Greek words rhegos, meaning blanket, and litho, meaning rock. Merrill wrote the following in 1897.

> In places this covering is made up of material originating through rock-weathering or plant growth in situ. In other instances it is of fragmental and more or less decomposed matter drifted by wind, water or ice from other sources. This entire mantle of unconsolidated material, whatever its nature of origin, it is proposed to call the *regolith*.

Water

Water plays at least three major roles in chemical rock weathering. First of all, water itself is an active weathering agent. In addition, the other main agents of chemical weathering, oxygen and carbon dioxide, can attack the rock structure only when dissolved in water. Finally, water carries away the products of weathering, allowing the process to continue. The first and most obvious thing that water does is to make things wet, a process that scientists refer to as hydration. Almost anywhere on Earth, except for the dry deserts, the soil and the underlying regolith are moist, often to a depth of several meters or more. This is important because having the regolith hydrated means that rock surfaces are coated with water films and rock weathering is a "wet chemistry" process. Table 3.7 compares the water content of weathered rock near the surface with that of relatively unaltered or intact rock below. The granite data are from Goldich (1938) and the diabase data are from Harrison (1933). Diabase, like granite, is an igneous rock, but it is finer grained than granite and contains less silica.

Table 3.7. Water content of intact and weathered rock (number of H_2O molecules in a rock cell containing 160 oxygen atoms)

Rock type	Intact rock	Weathered rock
Diabase	10.76	18.29
Granite	3.04	27.42

A water molecule is formed when two hydrogen atoms link up with a single oxygen atom. Each hydrogen atom shares its electron with oxygen, enabling the oxygen atom to gain the two electrons it needs to fill its outer shell. The result is a water molecule shaped much like a baby panda's face, with the large oxygen ion forming the face and the two hydrogen bumps forming the ears. The oxygen atom (the face), having taken on two electrons, has a slight negative charge, while the hydrogen atoms (the ears) have a slightly positive charge. The result is a polar molecule, with negative and positive electrical charges on opposite ends.

Some find it hard to believe that water is such a powerful weathering agent, because we intuitively think of it as a bland, neutral liquid. But water is far from inert; according to Keller (1957) water is "a powerful chemical reagent always supplying active H and OH ions. To feldspar and other rock-forming silicates, water is probably public enemy number one." The positive and negative ends of the water molecule exert strong forces that can tear apart other molecules, so substances such as sugar and salt are readily dissolved. Rocks are more resistant, but over millions of years, water has dissolved vast amounts of rock and washed untold numbers of atoms away to the sea.

Another important fact to remember is that there is no such thing as pure water. Regardless of the conditions, a few water molecules are always broken apart into hydrogen ions (H^+) and hydroxyl (OH^-) ions. The H^+ ion is essentially a proton with no electron attached, but an errant H^+ ion does not remain unattached for long; it quickly hitches a ride on a passing water molecule, creating an H_3O^+ ion. This tiny water-borne proton is a very effective weathering agent, easily making its way into rock structures and displacing positively charged ions such as Na^+, Ca^{2+}, and K^+. The two equations below illustrate the process of hydrolysis, showing how H^+ ions break down orthoclase, a kind of feldspar found in most igneous rocks:

$H^+ + H_2O \Rightarrow H_3O^+$

$KAlSi_3O_8 + 4H_3O^+ \Rightarrow K^+ + Al^{3+} + 3H_4SiO_4$

Hydrolysis requires water to proceed — and a single saturation is not enough. Geologists commonly find fresh rock with few signs of weathering at depths that are continuously saturated. In contrast, rocks closer to the surface, where the water table fluctuates up and down, display many signs of chemical weathering. Hydrolysis is favored not by continuous saturation with water that is immobile but instead by repeated infusions of fresh water to leach away the products of weathering as it passes through the rock. Hydrolysis has global implications, as noted by Keller (1957).

> It is interesting to note the over-all result of the world-wide hydrolysis of silicate rocks during geologic time. Much water has been dissociated into H and OH ions since the time when chemical weathering began. The OH ions have gone with metal cations to the ocean and have given rise to an alkaline ocean. The H ions have combined with alumina-silicate anions to form clay minerals, which are characterized chemically as being ... slightly dissociated (weak) acids. The land is the acid part of the reaction, the ocean the alkaline part.

Temperature

Temperature, unlike water, is not a direct agent of rock weathering, but it strongly influences both the intensity of weathering and the depth to which it occurs. According to Jenny (1941),

> A familiar observation of the early soil scientists was that in humid warm regions the rocks had weathered to much greater depths than in the cold zones. In northern Europe, in the Alps, and in the northern United States and in Canada, the thickness of the soil is usually expressed in inches or centimeters... In contrast, the weathered mantle of subtropical and tropical regions achieves huge thicknesses, and often one must dig for many feet or yards before the fresh rock is exposed. Depths of from 130 to 160 feet have been frequently observed, and Vageler reports a case of 1,312 feet.

Just knowing the mean annual temperature of an area, along with the amount of annual rainfall, usually enables one to surmise the rate of weathering as well as the direction and intensity of soil formation. Accordingly, temperature, rainfall, and the interactions between them are discussed more fully in subsequent chapters.

Carbon Dioxide

Water at a neutral pH of 7 contains equal amounts of H^+ and OH^- ions. Anything that increases the number of H^+ ions will make the water more acid and increase the rate of hydrolysis. One such thing is carbon dioxide. Some of the carbon dioxide from the air dissolves in rainwater to produce a dilute solution of carbonic acid:

$$H_2O + CO_2 = H_2CO_3$$

The H_2CO_3 then dissociates into H^+ and HCO_3^- ions, with the rate depending on pH and other conditions. Both the H^+ ions and the HCO_3^- ions play important roles in rock weathering. The H^+ displaces elements such as calcium and potassium, destabilizing rock structure, while the HCO_3^- acts as a balancing ion for the displaced cations, helping to keep them in solution; e.g., Na^+ and HCO_3^- ions can float away attached to different ends of water molecules. In general, every plus ion (+) carried away to the ocean is accompanied by a balancing negative ion (-), and most often that ion is bicarbonate (HCO_3^-). As a result, the world's rivers carry large amounts of dissolved HCO_3^- ions (typically listed as carbonate, or CO_3^{2-}). Some examples are listed below (from Clarke 1924). These data show that rock weathering, in addition to removing mineral elements such as calcium and sodium from rock, is continually removing carbon dioxide from the air and transporting it to the sea. This is an important point, one we will revisit in a later chapter.

Table 3.8. CO_3^{2} content of selected rivers (percent of total load)

River	CO_3^{2-} content
Chattahoochee River, Georgia	21.32
Amazon River, South America	34.75
Mississippi River, New Orleans	34.98
Shenandoah River, West Virginia	47.22

Oxygen

About 21 percent of the atmosphere is made up of oxygen (O_2) molecules. As discussed earlier, the oxygen atom has a total of eight electrons, two in the first shell and six in the second (2, 6), so it is always on the lookout for two more in order to reach the magic number of ten (2, 8). Atmospheric oxygen is so desperate for electrons that individual atoms get together and share electrons similar to the way two hikers lost in the woods might pool their meager rations. Each atom in the pair partially claims two electrons of the partner atom as its own, a less than perfect solution. As a result, atmospheric oxygen is always on the lookout for a more suitable liaison, and iron is perhaps the ideal alternative. When iron meets oxygen, a rapid exchange of electrons usually ensues, and this is not good news for iron-bearing minerals, as in the example below (Keller 1957).

$$4 \ FeSiO_3 + O_2 \Rightarrow 2 \ Fe_2O_3 + 4 \ SiO_2 + 512 \ kilocalories$$

In this reaction, iron within the silicate structure is oxidized to Fe3$^+$, liberating a significant amount of energy in the process. The mineral structure is disrupted by this reaction, creating iron oxide and silica, with some of the silica subject to removal by water. Whenever the iron in a mineral reacts with oxygen, the mineral structure starts to fall apart; so, not surprisingly, iron-bearing rocks are among the first to show the effects of weathering. Although oxidation is carried out by atmospheric oxygen, one should not minimize the role of water. The oxidation of minerals by atmospheric oxygen occurs almost entirely by the intermediate action of water, which can be present in quantities ranging from thin films of moisture to complete saturation. Water with a pH of 7 exposed to ordinary air develops an oxidizing potential of 0.81 volt, more than enough to oxidize Fe^{2+} to Fe^{3+} (Mason 1952).

Weathering and Time

In the reaction above, four moles of iron (4 Fe) are oxidized and 512 kilocalories of energy are produced. A mole of an element is simply its atomic weight expressed in grams. The oxygen atom, for example, has an atomic weight of 16, so a mole of oxygen weighs 16 grams. Now let us turn to iron. The atomic weight of iron to the nearest whole number is 56, so a mole of iron weighs 56 grams and four moles of iron (4 x 56) weigh 224 grams, or approximately half a pound. In the equation above, 512 kilocalories of energy are released for every half pound or so of iron that combines with oxygen during rock weathering. This is about the same number of calories you or I might consume in a typical lunch. The energy of oxidation and nearly all other reactions related to chemical rock weathering are similar to human metabolism in terms of energy intensity.

> Whereas activities such as mountain building and the reshaping of continents by glaciers involve spectacular amounts of energy... weathering has plodded along at a few thousand calories per mole throughout the ages, but it has nevertheless carried along organic evolution as a by-product (Keller 1957).

The Kamenetz Fortress

Considering the relatively low levels of energy involved in chemical rock weathering, how long does it take for rock to turn into soil? Jenny (1941) tells the story of the Kamenetz fortress, which was built in 1362 in Ukraine, occupied until 1699 and then abandoned for the next 230 years. In 1930, a Russian scientist studied a soil profile that had formed on one of the tower walls of the old fortress. He discovered that a layer of soil about twelve inches (30 centimeters) thick had formed during the 230 years the fortress had been vacant, or about five inches (13 centimeters) per century (Akimtzev 1932). At this rate it would take about 800 years for a soil one meter deep (about 40 inches) to form.

This is in line with other estimates; forming soil from solid rock can be a very slow process, because physical weathering and chemical weathering have to take place concurrently. Fortunately, that does not happen too often. Most often, weathering and soil formation take place in rock material that has already been broken down into small particles. In contrast to the hundreds of years that might be needed with solid rock, a soil can form in material such as glacial till, loess, river alluvium, or marine sediments in only a few decades. Think of the Midwestern farm belt, the Palouse region of the northwest, the Central Valley of California, the Mississippi Valley, the Atlantic Coastal Plain, and the Red River Valley—all are major agricultural regions of America where soil formation took place in rock material that had already been broken down into small particles by wind, water, or ice. In such areas, it would have taken only a few decades for fully functional soils, both biologically and chemically, to come into existence.

This chapter has covered a number of fairly complex topics, so before continuing, it might be helpful to review some of the main points.

1. A surprisingly small number of chemical elements account for nearly 99 percent of the weight of rocks at the Earth's surface. These are oxygen, silicon, aluminum, iron, calcium, magnesium, potassium, and sodium— the "big eight." Seven of the big eight elements form ionic bonds with

the eighth, oxygen, by donating a number of electrons sufficient to reach "magic number" status and become stable. The strength of these bonds varies widely, but all are of about the same magnitude as human metabolism. For example, the approximate bonding energy of K_2O (299 kcal/mole) is about equal to the food energy in one fast food hamburger, while that of SiO_2 (3110 kcal/mole) is about the same number of calories a man performing physical labor in a cold climate might consume daily.

2. Water, carbon dioxide, and oxygen are the three main agents of rock weathering. When it comes to silicate rocks, the water that we view as bland and innocuous is public enemy number one. It works slowly but unceasingly to pry the elements from rocks and then carry them away to the sea. In addition, carbon dioxide and oxygen can only attack and break down rock when dissolved in water. Processes of chemical weathering involve relatively small amounts of energy — about the same magnitude as human nutrition, so creating a soil from solid rock can take many hundreds of years. But in finely divided rock material, such as loess or river alluvium, where the bulk of physical weathering has already been done, soil can form in only a few decades.

3. Although not an active agent of weathering, temperature strongly affects the activities of water, carbon dioxide and oxygen in breaking down rocks. For example, the deep soils and deep weathering zones common to the tropics require a combination of nearly constant high temperatures and large amounts of rainfall.

4. Sodium was determined to be the most mobile of the rock-forming elements and was assigned a relative mobility of 100. The elements ranked as to relative mobility are sodium (100), calcium (45), magnesium (30), and potassium (20), followed by silicon, iron and aluminum, all of which are at least 20 times less mobile than sodium. These differences in mobility strongly suggest that the elements take very divergent paths after being pried from the rock structure.

The next chapter takes a close look at silicon, aluminum, and iron, the least mobile of the rock-forming elements. Why are they more resistant to weathering and leaching than the other elements in rock? More specifically, what forms do they assume and what transformations do they undergo as rock is transformed into soil? In a broader sense, what use does nature make of them?

CHAPTER 4. CLAY AND HOW IT FORMS

Mineral soil is made up of sand, silt, and clay. Sand grains are big enough to be seen by the human eye and persist for a long time in soil because the quartz (SiO_2) of which they usually are made is virtually indestructible. SiO_2 can be broken down in a laboratory, but doing so is rather pointless, since the silicon and oxygen will quickly recombine to form quartz again. Being largely inert, sand contributes little to the nutrition of plants; but it has other virtues. Large, angular sand particles do not fit closely together, so they create voids which, though small individually, result in a lot of aggregate space, and this is important. In an apparently solid clod of earth, only about half of the volume is truly solid; the other half is empty space. Of course, this space is never really empty; the larger spaces are filled with air, while the smaller pores and voids hold water against the downward pull of gravity. These air and water filled spaces also make it possible for roots to penetrate the soil and they provide a snug home for the many life forms that thrive in the soil. Without large, irregular sand grains to help create open spaces, soil would lose much of its biological value.

The largest silt particles are 50 microns in diameter, about half the thickness of a human hair, not quite visible to the human eye. A handful of silt feels smooth and floury to the touch. Most silt particles are quartz, so like sand grains they contribute relatively little to the nutrition of plants. But silt, like sand, is important in determining the physical properties of soil. An optimal amount of silt enables soils to store more water and make more of it available to plants. Some silt particles contain minerals such as feldspar and mica, which upon weathering release nutrients into the soil. It is no accident that some of our most productive agricultural soils, in places such as Iowa and Illinois, are high in silt.

Now we come to clay, which is the source of much of the chemical activity in soil. The largest clay particles are 2.0 microns in diameter, about 1/50th the thickness of a human hair. Because the individual particles are so small, clay is sticky when wet and becomes very hard when dry; it swells when wet and shrinks upon drying. When stirred into pure water, clay clouds the liquid and can remain in suspension for many days; and unlike sand or silt, clay is very active chemically. Chemical activity is perhaps the most important attribute of clay, enabling it to store mineral nutrients and make them available to plant roots. An astoundingly large surface area enhances the chemical activity of clays. A single ounce of clay taken from a field at Rothamsted, England, site of the world's oldest agricultural experiment station, had a surface area of nearly 2.5 hectares, or more than six acres. Chemical reactivity and very high surface area account for many of the properties that make soil clay so important in terrestrial ecosystems.

Humans have had a long association with clay. In order to settle permanently in villages and towns, people needed a way to transport and store agricultural products such as grain, wine, and olive oil; and in many societies, clay made that possible. Clay pottery was the first truly synthetic product to be made and this technology, like agriculture itself, appeared independently in many parts of the world. Pottery fragments are the most common artifacts found when archaeologists dig up ancient settlements. Here is an interesting description of the ancient city of Uruk from the Sumerian epic *Gilgamesh*, written more than 2,000 years ago: "One part is city, one part orchard, and one part claypits. Three parts including the claypits make up Uruk."

Centuries of pottery and brick making taught us much about the physical properties and behavior of clay. But early farmers saw clay from a different perspective than potters and brick makers, observing how varying amounts of clay in the soil affected the usefulness of land for agriculture. They learned to avoid sandy, droughty soils that were too low in clay as well as heavy, sticky soils that were too high in clay, instead seeking out loamy, light-textured soils commonly found along rivers and streams. Early farmers also learned how to modify soils that contained either too much or too little clay. For example, England has many areas of sandy soils on barren heaths and moorlands that cannot hold enough water to grow crops unless improved. Some of these areas were turned into productive farms during the 1700s and 1800s by mixing clay mined from nearby areas, often as much as 100 tons per acre, into the surface.

But just as some areas of England were not clayey enough, other areas contained too much clay, making the land difficult to farm; early Britons found a solution to this problem also. They learned, perhaps from the Celts, that adding large amounts of chalk ($CaCO_3$) made a clay soil much less sticky and gave it a friable structure, allowing water to move into and through the soil more easily and making cultivation much easier. This method was used even before the Romans came to the island, as described by Pliny in his *Natural History* (AD 79).

> The peoples of Britain and Gaul have discovered another method for nourishing the land. There is something they call marl... It is sought for deep in the ground, wells frequently sunk one hundred feet deep... It lasts for eighty years and there is no evidence of anyone who has put it on twice in his lifetime.

Until well into the 19th century, it was common for anyone acquiring land with heavy clay soils, especially in the eastern and southern counties of England, to "chalk" it heavily enough to last a lifetime. According to Sir John Russell (1957), long-time director of England's Rothamsted Experiment Station,

> The method [chalking the land] remained in use right up to the time of the First World War; it was a traditional craft carried out by gangs of itinerant workmen who were very skillful at finding the best points for sinking the wells... In 1915 I engaged a gang to chalk one of the Rothamsted fields; they did it exactly as Pliny had described.

All or even most of the clayey soils in Great Britain, however, could not be improved with chalking, nor could most of the sandy soils be improved by adding clay. So well into the 20th century, the amount of clay in the soils of a region largely determined the nature and profitability of agriculture as well as many other aspects of British life. In his 1911 book, *Lessons on Soil*, E.J. Russell offered a number of fascinating insights into the effects of soil clay on British life. In the next few pages, we quote extensively from this work, which, although more than a century old, is an excellent book on soil science. Prior to about 1870, clay soils were used extensively to grow wheat and other crops in England despite the fact that they were relatively unproductive and very difficult to work. Russell wrote, "We have seen that clay holds water and is very wet and sticky in winter, while in summer it becomes hard and dry, and is liable to crack badly... But in days when we grew our own wheat, before we imported it from the United States and other countries [prior to the late 1800s], this clay land was widely cultivated for wheat and beans."

Russell then explained what happened to the clay districts as increasing amounts of wheat were imported from the United States and Canada. "Land either went out of cultivation like the 'derelict' farms of Essex, or it was changed to grassland and used for cattle grazing. Great was the distress that followed; some districts indeed were years in recovering."

In addition to their limitations for agriculture, the amount of clay in the soil also affected settlement patterns, as clayey areas were avoided as long as possible. "Large tracts of clay which remain wet and sticky during a good part of the year are not attractive to live in, and even near London they were the last to be populated."

Soils with too much clay can cause problems, but so can sandy soils with too little clay. Here is how Russell characterized the poor, sandy soils of the English heathlands. "There were practically no villages and few cottages, because the land was too barren to produce enough food; the few dwellers of the heath, or the "heathen," were so ignorant and benighted that the name came to stand generally for all such people... As there were so few inhabitants the heath used to be great places for robbery, highwaymen, and evildoers generally."

Despite the danger from robbers and from the heathen in general, traveling through the English heaths had its attractions; quoting Russell again, "It is easy to travel in sand country because the roads dry very quickly after rain, though they may be dusty in summer. Sometimes the lanes are sunk rather deeply in the soft sand, forming very pretty banks on either side."

Some English soils were too clayey while some were too sandy, but Russell described another group of soils with just the right amount of clay, at least for agriculture. "Loams are well suited to our ordinary farm crops...the farmer on a good loam is in the fortunate position of being able to produce almost anything he finds most profitable...the farm houses and buildings are well kept, and there is a general air of prosperity all round."

Narratives such as these from Britain could just as easily have come from many other parts of the world where people lived close to the land for many generations. Long and intimate association with the land taught us a lot of practical things about clay, but we did not begin to unravel its true, or rather, its scientific nature until well into the 20th century. In 1887, Le Chatelier proposed what came to be called the "clay mineral concept." This is the idea that there are only a limited number of clay minerals and that all of them are made up of small crystals. X-ray diffraction studies later showed that Le Chatelier had been correct; soil clays are indeed crystalline and every sample analyzed is made up of the same small group of minerals. Beginning about 1924, scientists at the United States Geological Survey began a series of studies on the structure of clay minerals that lasted more than two decades. C. S Ross was a leader in these efforts, and he and his group did much to expand our knowledge, publishing such papers as *The Mineralogy of Clays* in 1927, *The Kaolin Minerals* in 1931 and *Minerals of the Montmorillonite Group* in 1945. As a result, the clay mineral concept had been firmly established by the middle of the 20th century and the structure and chemical composition of most clay minerals had been determined.

By the early part of the 20th century, researchers had learned that clay minerals are composed primarily of three elements — silicon, aluminum and oxygen. The early work of Linus Pauling provided the basic concepts needed to understand why it is mostly these three elements that form clay minerals. We discussed Pauling's rules earlier, pointing out their usefulness in understanding rock structure; they are equally useful in understanding the formation and structure of clays. Pauling's 1929 paper, *The Principles Determining the Structure of Complex Ionic Crystals*, is especially relevant. This paper explains why the two basic building blocks of soil clays are the silica tetrahedron and the aluminum octahedron.

The Silica Tetrahedron

The silica tetrahedron consists of a silicon ion (Si^{4+}) enclosed by four oxygen ions, creating a tiny triangular pyramid. The silicon fits snugly within the small opening created by the four large oxygen ions. Although the pyramid model is useful, a more appropriate image might come to mind when one considers the relative sizes of the ions involved. An oxygen ion is more than three times larger than a silicon ion, so instead of a sharp-angled pyramid, a more accurate image might be of four soccer balls enclosing a tennis ball. You would still have the pyramidal shape, just not the sharp angles shown in so many diagrams of the tetrahedron. There are some excellent drawings and models of this structure online. You can access them by keying a phrase such as "silicon tetrahedron pictures" into a search engine.

The Aluminum Octahedron

The geometric form of an octahedron, an eight-sided geometric structure, is harder to visualize than a tetrahedron. To do so, first imagine a pyramid with a base having four corners (a square) coming to a single point at the top. This creates a pyramid with a square base and four triangular faces. Now visualize an identical pyramid sitting beside the first one. If you mentally pick up both pyramids and join their bases together, you will have formed an octahedron with an oxygen ion at each of the six corners.

Now all you have to do is imagine a single aluminum ion at the center of the octahedron. If you find it too difficult to form these images in your mind, there is a good alternative. Simply key the phrase "aluminum octahedron pictures" or something similar into a search engine and you will be directed almost instantly to a variety of diagrams depicting the basic structure of the aluminum octahedron.

Why are these two geometric structures the principal building blocks of clay minerals? Why did nature not choose some other geometric structure or some other elements? A set of rules developed by Linus Pauling early in the 20th century will help us answer these questions. Since we are interested in clay minerals, the discussion will be specific to crystals with oxygen as the enclosing ion. Pauling's first rule, the radius ratio rule, says simply that the smaller a central ion is, the fewer oxygen ions it will surround itself with. The radius ratio is calculated by dividing the radius of the enclosed ion by the radius of the enclosing ion, which in our case is oxygen. Using Pauling's rule, we can predict whether an ion will surround itself with four oxygen ions, six oxygen ions, eight oxygen ions, or even more.

Silicon is a good example. According to Pauling's first rule, the required radius ratio for an ion to be surrounded by four oxygen ions ranges from 0.22 to 0.41. The silicon ion has a radius of 0.42 Å. If you divide 0.42 Å by the radius of the oxygen ion (1.32 Å), the answer is 0.32, right in the middle of this range; so the fact that the silicon ion is enclosed by four oxygen ions in the crystal structure is in accordance with Pauling's rules. Now consider the aluminum ion, which is normally surrounded by six oxygens to form an octahedral structure. According to Pauling, the radius ratio for a central ion to be surrounded by six oxygen ions ranges from 0.41 to 0.73. The radius of the Al^{3+} ion is 0.51 Å. If you divide this by the radius of the oxygen ion (0.51/1.32), the answer is 0.39. This places Al^{3+} at the boundary between four oxygen ions and six oxygen ions. Although aluminum usually surrounds itself with six oxygen ions, a significant number end up enclosed by four oxygen ions, resulting in some tetrahedral building blocks with aluminum as the central ion instead of silicon.

Although there is some replacement by other ions, most tetrahedral units have silicon as the central ion and most octahedral units have aluminum as the central ion. One reason for this is the sheer abundance of silicon and aluminum in the world. Another reason has to do with a second dictum laid down by Pauling, the electrostatic valence rule. This rule states that the relative strength of the bond between a central ion and the oxygen ions surrounding it can be estimated by dividing the valence of the enclosed ion by the total number of enclosing oxygen ions. Valence, for our purposes, is simply the number of electrons the central ion wants to give away. The valence of the silicon ion, for example, is 4 and it is enclosed by four oxygen ions, so the relative bond strength is 4/4 = 1.00. The relative bond strengths for all of the major ions involved in the structure of clay minerals are shown below.

Table 4.1. Relative bond strength between various ions and oxygen in clay minerals

Ion	Relative bond strength
Si^{4+}	1.00
Al^{3+} (enclosed by 4 O^{-2} ions)	0.75
Al^{3+} (enclosed by 6 O^{-2} ions)	0.50
Mg^{2+}	0.33
Fe^{2+}	0.33

Not only is silicon the right size to fit in the niche created by four oxygen ions, its high charge (4^+) means that the bond with oxygen is very strong. Aluminum will fit into the hole created by four oxygen ions, but it is not as good a fit as silicon and, in addition, aluminum has a weaker bond with oxygen. As a result, silicon is the ion of choice for the tetrahedral unit in clay crystals. Based on size alone, aluminum is not the best fit for the opening created by six surrounding oxygens, but it benefits from being very abundant and from having a relatively strong bond with oxygen. The slightly larger Mg^{2+} ion, with a radius of 0.66 Å., is actually a better fit. But there are a lot more Al^{3+} than Mg^{2+} ions in the world and, in addition, the Al-O bond is much stronger than the Mg-O bond, as per Pauling's electrostatic valence rule. As a result, aluminum is the ion of choice for the octahedral units in clay minerals. It is no surprise that most clay minerals consist primarily of aluminum, silicon and oxygen, and are properly referred to as aluminosilicates.

Crystal Structure

Holding a handful of moist soil, it is hard to imagine that the clay making it feel sticky actually consists of billions of tiny crystals, but it does; and like the crystals you might have grown for a science project in your youth, these crystals grow in the soil. Hazen and Trefil (2009) characterized crystals as "regular three-dimensional arrays of atoms, repeated over and over again in a sort of Tinkertoy arrangement...something like a huge stack of boxes, with each box the same size and shape and all the boxes holding exactly the same atomic contents." Each "box" in a crystal is commonly referred to as the "unit cell," and crystals are constructed by joining thousands or even millions of them together. Once you know the makeup of a unit cell, you know a lot about the crystal. In soil, tetrahedral units and octahedral units organize themselves to produce two basic kinds of unit cells, those with a 1:1 structure (Si-Al) and those with a 2:1 structure (Si-Al-Si). Accordingly, soil clays, which are built from many millions of these unit cells, usually are designated as either 1:1 or 2:1 clays. Montmorillonite, a 2:1 clay, and kaolinite, a 1:1 clay, both have a sheet structure, because they can grow laterally by adding more crystal units to the edges, but they cannot grow vertically.

In kaolinite and similar 1:1 clays, the top surface of the sheet, which is made up of silica tetrahedra, has oxygen ions exposed, while the bottom, the aluminum octahedral layer, has hydroxyls (OH) exposed. This allows the bottom of one sheet to attach itself to the top of another sheet rather firmly by forming hydrogen bonds. The individual bonds between the oxygen and hydroxyl ions are weak, but there are so many of them that the sheets are held closely together with enough force that water molecules cannot easily squeeze between the layers. This gives kaolinite and other 1:1 clays one of their most important properties. They expand very little when wet and do not contract much when dried out. Low shrink-swell, as this is called, is the main reason 1:1 clays are often suitable for making pottery or bricks, and the reason the walls and foundations of structures built on soils with 1:1 clays are less subject to cracking.

The 2:1 Clays

Montmorillonite is the best-known member of the 2:1 group of clays. The name comes from the French town of Montmorillon, near the location where one of the earliest examples of this mineral was found. According to Grim (1953), montmorillonite has a kind of sandwich structure, with a sheet of aluminum octahedra sandwiched between two silica tetrahedral sheets (Si-Al-Si). Grim's iconic 1953 text included some excellent drawings of both kaolinite and montmorillonite structures. I was able to access and view these drawings online by keying "pictures of clay structure Grim" into a search engine. It might be helpful to view drawings of these structures online before proceeding.

Sheets of montmorillonite can stack up in the soil much like kaolinite does, but the adjacent sheets of montmorillonite are not held together by hydrogen bonds. Because of the sandwich arrangement (Si-Al-Si), the bottom of one sheet consisting of silica units is always touching the top of another sheet also consisting of silica units, so there is little or no chemical attraction between them (i.e., Si-Al-Si : Si-Al-Si). This accounts for one of the most distinctive and notorious features of some 2:1 clays. Since there is little or no chemical bonding between the layers, water molecules can move freely in and out, causing the clays to expand when wet and to shrink again upon drying out. Soils that contain such clays can pose significant problems. Below is an excerpt from a 1987 report issued by the U.S. National Academy of Sciences titled "Confronting Natural Disasters."

> Expansive soils — soils that exhibit large potential for shrinking and swelling with changes in moisture content — are another long-term hazard. Construction on these soils is extremely vulnerable to damage — even total destruction — as the ground surface elevation changes in response to seasonal fluctuations in rainfall. The problem is particularly acute in arid and semiarid regions... The total cost of damage associated with expansive soils is estimated at a minimum of $6 billion per year in the United States alone; it is the nation's most costly natural hazard.

The Chemistry of Clay

Another interesting and important thing to know about clay is that even a small amount has a huge surface area and that these surfaces have millions upon millions of negative charges. For example, just one teaspoon of almost any clay contains at least one billion billon negative charges scattered across its surfaces. That is the number one followed by 18 zeros. Since nature does not tolerate charge imbalances, clay surfaces attract and hold positively charged ions such as Ca^2, Mg^{2+}, K^+, and H^+ to even things out. These attached or "exchangeable" ions serve as a nutrient supply for plant roots and soil microorganisms.

The total number of electrical charges a given amount of clay has on its surfaces is referred to as its cation exchange capacity (CEC). Since CEC is an important indicator of quality, soil chemists have studied it intensely. One thing they have discovered is that the CEC of 2:1 clays is usually ten or more times higher than that of 1:1 clays. Here are typical values in milli-equivalents/100 grams of clay.

Kaolinite (1:1 clay): 3–15

Montmorillonite (2:1 clay): 80–150

If you read the most recent soil science literature, you will see clay activity expressed in centimoles per kilogram, which is in accordance with the International System of Units or SI. This book employs the older unit, milliequivalents per 100 grams, for two reasons. First, it has a very long history in the soil science literature, and second, it is very useful in explaining some important concepts. The conversion is a simple one; the two units are exactly the same numerically.

Soil chemists originally chose milliequivalents per 100 grams as the unit for measuring clay activity largely because of convenience. The resulting numbers are very manageable and easy to use in calculations, being neither too large nor too small. For our purposes, it is not necessary to understand exactly what these units mean or how they were derived, except to note that each milliequivalent equates to 6.022×10^{20} electrical charges in the soil. This means that even a small amount of kaolinite has an astounding number of surface charges. The maximum cation exchange capacity listed for kaolinite above (15 milliequivalents/100 grams) equates to many billions of electrical charges on the surfaces in as little as a tablespoon of clay. These numbers are so large because clays are truly electrified. When Walt Whitman wrote *I Sing the Body Electric*, he could have been writing about soil.

But when it comes to electrical charge, kaolinite pales in comparison to montmorillonite. The following thought experiment illustrates this. First you must, at least in your mind, bury the entire state of Texas under a layer of sand 50 feet (15 meters) thick. This should be "fine" sand, meaning that each grain should be about 1/100th inch (0.25 millimeter or smaller) in diameter. The next step is to count every single grain and record the total. According to Bob Everson of Purdue University, who is credited with designing this exercise, the total number of sand grains should equal about 6.022×10^{23}, the famous Avogadro's number, which is the number of atoms in a mole of any element. This very large figure also happens to be the number of electrical charges in an "equivalent." Soil chemists, of course, measure cation exchange capacity in milliequivalents, which is 1/1000th of an equivalent, so each milliequivalent measured in soil is equal to 6.022×10^{20} electrical charges. Since we are dealing with milliequivalents instead of equivalents, we had to knock three zeros off of Avogadro's number.

What quantity of soil would have a total number of surface charges equal to 6.022×10^{23} or Avogadro's number? That depends on whether the clays in the soil are 1:1 or 2:1. It also depends on how much organic matter is in the soil, but that is a topic for a later chapter. Assume that two soils are sampled, one from a field in Iowa and the other from a field in Georgia. Let us further assume that all of the clay in the Iowa soil is montmorillonite, while all of the clay in the Georgia soil is kaolinite. This is not an unreasonable scenario, since kaolinite is very common in Georgia and montmorillonite is very common in Iowa. If you went to a typical cornfield in Iowa, dug up just ten pounds or so of surface soil and carried it out of the field, you would be walking away with at least 6.022×10^{23} negative charges, or one equivalent. If you then turned each of these charges into a sand grain, you would have conjured up enough sand to cover the entire state of Texas with a layer of sand 50 feet (15 meters) thick. But in order to harvest the same number of charges from a typical cornfield in Georgia, you would have to carry ten times as much soil (about 100 pounds) out of the field. Note: These calculations assume that in both cases the soils contain about 20 percent clay by weight.

The fact that a typical Iowa soil, dominated by 2:1 clay has so many more surface charges than a similar soil in Georgia dominated by 1:1 clay has important implications for agriculture. This is because every negative charge on a clay surface is balanced by an attached ion, and the most common ones are calcium, magnesium, and potassium along with quantities of hydrogen and aluminum, which vary depending on the acidity of the soil. Calcium, magnesium and potassium are important mineral nutrients, so having an abundant supply of them in the soil is important for crop production. The Iowa soil is capable of holding many times more mineral nutrients in exchangeable form (meaning that they are available for plant uptake on fairly short notice) than the Georgia soil. Assume that the Iowa soil contains 15 percent by weight montmorillonite clay in the surface layer, while the Georgia soil contains 15 percent by weight kaolinite clay. It is fairly easy to calculate the amounts of Ca, Mg or K that can be held by the negative charges in the surface layer (the top 15 cm or about 6 inches) of each soil at any given time.

If we assume that Ca^{2+} ions occupy about one third of the charges on clay surfaces, some straightforward calculations show that a soil surface containing 15 percent clay could hold about 200 pounds of Ca per acre if all of the clay is of the 1:1 type. In contrast, a soil surface with the same amount of 2:1 clay will hold about 2,000 pounds of Ca per acre. It all comes back to the simple fact that the average CEC of 2:1 clay is about ten times greater than the average CEC of 1:1 clay, resulting in ten times more negative charges on the clay surfaces. It is easy to see how important the nature of clay is to agricultural production. As a rule, soils with 2:1 clays are much more fertile and typically produce much higher crop yields than similar soils with 1:1 clays.

But the significance of cation exchange goes well beyond agriculture. The development of electrical charge on clay surfaces is one of the most important phenomena in nature. Had this not happened, many familiar ecosystems would be missing from our world; it is even possible that our species would not have evolved. It is almost certain that human society as we know it would not exist. Without electrically charged clays to hold ions in places, elements such as calcium and magnesium would be leached from the land surface and carried away to the sea soon after being freed from rocks, creating "mineral deserts" in areas of higher rainfall.

The ions stored on clay surfaces are like money in a checking account, a source of nutrients from which the plant roots can make frequent with-drawals; and like a checking account, the "funds" can be replenished at intervals from the breakdown of previously un-weathered rock fragments (such as mica) in the soil or from added fertilizer. It is clear that a typical Iowa soil with 2:1 clays has a much larger pool of readily available nutrients than a typical Georgia soil with 1:1 clays. The fact that Georgia soils gener-ally have less exchange capacity than Iowa soils does not mean farming cannot be productive in Georgia. There are many successful farmers there and agriculture is an important part of the state's economy. But one statistic is telling. In 2014, less than ten percent of the land area in Georgia was used to grow crops, while the comparable figure for Iowa was nearly 70 percent (National Agricultural Statistics Service 2015).

We did not have to pick on Iowa and Georgia. Instead of going to Iowa for our hypothetical soil sample, we could have gone to any number of nearby states such as Illinois, Minnesota or Missouri. Similarly, instead of going to Georgia, we could have gone to North Carolina, South Carolina, Alabama, or even Virginia. But we made good choices; Iowa is in the heart of "2:1 clay country" and Georgia is in the heart of "1:1 clay country." This is because there is a strong correlation between regional climate (which generally means rainfall and temperature) and the kinds of clay that form in soils. There are exceptions of course, most often due to unusual rock types, but broad areas of the United States have similar kinds of geology, rainfall and temperature; and as a result, distinctive kinds of clays that dominate their soils.

Climate and Clay

The balance between annual rainfall and annual evapotranspiration is a strong predictor of the amount and kinds of clay that form in soils. We will examine the interactions between temperature and rainfall and how they affect clay formation in four very different areas of the United States:

1. Desert soils of the Southwest, where evapotranspiration greatly exceeds rainfall every month of the year.

2. Former grassland soils of the Great Plains, where rainfall and evapotranspiration are about equal throughout the year.

3. Forested soils of the Southeast, where rainfall exceeds evapotranspiration during the winter, but is slightly less than or equal to evapotranspiration during the summer.

4. Forested soils of the Northeast, where rainfall exceeds evapotranspiration in both winter and summer.

The American Desert

Deserts generally receive fewer than ten inches (25 centimeters) of annual rainfall and most of that quickly evaporates, so the soil is wetted to only a shallow depth if at all. As a result, leaching intensity is very low. According to Jenny (1941), "In arid regions, all rainwater that penetrates into the soil is either held by the soil particles or moves upward again through evaporation and transpiration by plants. The products of weathering processes are not removed from the soil through leaching."

It is a characteristic of desert soils that nearly all of the elements released by chemical weathering stay on site. As a result, there is no shortage of silicon and no shortage of magnesium, so the clays produced are overwhelmingly of the 2:1 type. An abundant supply of silicon favors the production of 2:1 clays because there are about two silicon atoms for every aluminum atom in clays such as montmorillonite; so building 2:1 clays requires a lot of silicon. In addition to silicon, an adequate supply of magnesium is needed to replace about one in every six or so aluminum ions in the crystal structure. This replacement is necessary to create the large number of surface charges typical of 2:1 clays. Remember, every time an Mg^{2+} ion replaces an Al^{3+}, there is a net gain of one negative charge on the clay surface.

Below are some excerpts from the official series description of Casa Grande, the state soil of Arizona. The Natural Resources Conservation Service (NRCS) maintains a database with a detailed description of every soil series classified and mapped in the United States. You can access this database by keying the phrase "official soil series descriptions" into a search engine.

Casa Grande Series

0 to 1 inch (0 to 2.5 cm): Fine sandy loam; violently effervescent; strongly alkaline (pH 8.6); slightly saline; slightly sodic.

1 to 5 inches (2.5 to 13 cm): Fine sandy loam; strongly effervescent; strongly alkaline (pH 8.6); very slightly saline; moderately sodic.

5 to 8 inches (13 to 20 cm): Sandy clay loam; soft accumulations of carbonates and salt; violently effervescent; very strongly alkaline (pH 9.6); moderately saline; strongly sodic.

This description continues on to a depth of 60 inches (1.5 meters) but the top eight inches (20 centimeters) are sufficient for our purposes. Words such as "alkaline, saline, sodic" and phrases such as "soft accumulations of carbonates and salt" keep saying the same thing: salty, salty, salty. The fact that very leachable salts are found so near the surface in desert soils such as Casa Grande makes it unlikely that more resistant ions such as silicon and magnesium could have been leached away. Clay production in desert soils is limited by very low rainfall, but low rainfall and lack of leaching also mean there is no shortage of Mg^{2+} and Si^{4+} ions, so any clay formed is typically of the 2:1 variety.

The Former Grasslands

Many people think of Denver, Colorado, as a cold and snowy place, but the city gets surprisingly little precipitation, slightly more than 15 inches (38 centimeters) per year. This is because Denver is in the rain shadow of the Rocky Mountains. The amount of annual precipitation increases in a regular, predictable fashion as you move away from Denver and go east across Colorado, through Kansas and on into Missouri. The regular increase in rainfall with distance makes this region ideal for assessing the effects of rainfall on soil formation.

Hans Jenny and his colleagues chose the uniform loess belt that extends through northern Kansas and into Missouri to study the effect of increasing rainfall on leaching and clay formation. Western Kansas gets about 15 inches (38 centimeters) of precipitation annually, while central Missouri, 500 miles to the east, gets more than 40 inches or about one meter. The 38 centimeters or so of precipitation that falls in western Kansas does not seem like much, but it is nearly twice the amount that falls in the deserts of the Southwest. Because of higher rainfall and cooler temperatures, the soils of western Kansas lack many of the salts found in the desert soils; but deposits of carbonates ($CaCO_3$ and $MgCO_3$) can still be observed within the first few feet of the surface. For example, the official series description of Harney, the state soil of Kansas, notes the presence of "many soft accumulations of carbonates" at a depth of less than 30 inches (76 centimeters).

But the farther east you go, the more leached the soils become. For example, the official description of Iowa's state soil, Tama, makes no mention of carbonates; the higher rainfall has leached them out of the profile. In the vicinity of eastern Nebraska and eastern Kansas, annual rainfall approaches 30 inches (around 75 centimeters), but annual evapotranspiration is still less than or equal to rainfall during most years. The relative amount of rainfall increases in Iowa and Illinois, but leaching is not excessive anywhere in the great loess belt that forms much of America's agricultural heartland.

Because rainfall and evapotranspiration are roughly in balance much of the year, the nutrient-rich parent material of the interior grasslands is subject to only a moderate degree of leaching; although much of the free sodium and potassium are leached away, significant amounts of calcium and magnesium remain, as well as most of the silicon. With an abundance of magnesium and silicon, it should surprise no one that 2:1 clays such as montmorillonite dominate the soils of this region. Quoting Keller (1957), "In this part of the country the rocks weather relatively rapidly to produce large amounts of the 3-layer, high ion-exchange type of clay."

Soils of the Southeast

This is what W.D. Keller (1957) had to say about the agricultural potential of soils in the southeastern United States.

> Here, where the soils are kaolinitic to lateritic, and the amount of organic matter is low, the cation exchange capacity of the soil is thereby reduced to low levels, and its productivity in terms of nutritive food is likewise low.

Professor Keller taught at the University of Missouri for many years, so his pronouncement concerning the low fertility of Southern soils might be a case of someone cheering for the home team — or rather the home soil. But he is not wrong; the soils in the Southeast lack the innate high fertility of Midwestern soils. He also is correct in his characterization of Southern soils as being "kaolinitic to lateritic." It has been well established that 1:1 clays such as kaolinite dominate the soils of the southeastern Piedmont and Coastal Plain and that there also are high levels of iron oxides (hence the term lateritic). The reasons for this are rather straightforward. Consider the average annual temperature in Des Moines, Iowa, compared to Athens, Georgia. The average annual temperature in Athens is 63° F, 12 degrees hotter than Des Moines, at 51° F. Athens also gets significantly more rain, which results in more leaching, mostly during the winter.

The chemical formula for kaolinite is $Al_2Si_2O_5(OH)_4$. Aside from oxygen and hydroxyl (OH) groups, kaolinite can consist entirely of aluminum and silicon. No magnesium or other ions are required in the structure, so it can form in environments where bases such as calcium and magnesium have been leached away soon after being freed from the rocks; and this is what happens in the hot, rainy Southeast. But the element to focus on here is silicon, and a comparison of two rivers, the Missouri and the Chattahoochee, will show why.

The Missouri, with a drainage area of nearly half a million square miles (1.3 million square kilometers), is the longest river in North America. From its origin near Three Forks, Montana, it travels more than 2,300 miles (3,700 kilometers) to St. Louis, Missouri, where it empties into the Mississippi. Its tributaries drain all of Nebraska and large portions of Montana, North Dakota, South Dakota and Wyoming. They also drain parts of Colorado, Kansas, Missouri, Iowa, and Minnesota. Much of this land, especially the western portion, is semiarid and dominated by soils high in 2:1 clays.

In contrast, the Chattahoochee is a relatively small river confined mostly to Georgia and Alabama. It starts out in northeast Georgia, flows past Atlanta and then makes its way to the Georgia–Alabama border, which it follows south for much of its length. Shortly before reaching the Florida border, it merges with the Flint River to form the Apalachicola, which then flows a short distance to the Gulf of Mexico. In contrast to the Missouri, the Chattahoochee drains an area that is markedly warmer and gets more rainfall, and as a result, both weathering intensity and leaching intensity are much higher. Decades of research and soil survey experience have taught us that many of the soils in Georgia and in much of the southeast are red and clayey, and that most of that clay is kaolinitic. The old phrase "red clay hills of Georgia" certainly applies.

The Chattahoochee and Missouri Rivers allow us to test the "predictability of nature" hypothesis introduced in Chapter 2. Soil clays in the Missouri River basin are predominantly of the montmorillonite type (Si-Al-Si), while most clay minerals in the Chattahoochee drainage area are of the kaolinite (Si-Al) type. But why is this so? Why are there so few

2:1 clay minerals in the soils of Georgia? Simple inspection tells us that the 2:1, or Si-Al-Si clays need to incorporate about twice as much Si in their structures as the 1:1, or Si-Al clays. River chemistry suggests that a lot more silicon is leached from Georgia landscapes than from the semiarid lands west of the Missouri River. Consider the numbers below from Clarke's Data of Geochemistry (1924).

Table 4.2. SiO$_2$ content (% of dissolved load) for the Missouri and Chattahoochee Rivers

Missouri	9.0
Chattahoochee	37.7

Silica accounts for almost 38 percent of the dissolved load in the Chattahoochee River compared with only nine percent in the Missouri. It is clear that much more SiO$_2$ is leached from the landscapes of hot, rainy Georgia than from the cooler, semiarid lands west of the Missouri River. This clearly reflects the kinds of clay formed in the two watersheds. In the Missouri basin, where much of the silicon stays on the landscape, mostly 2:1 clay minerals are formed; while in Georgia, where much of the silicon is leached from the land, nature makes mostly 1:1 clays.

The Northeastern Forest

As pointed out earlier, deserts are simply too dry for much clay formation to take place. The opposite situation exists in the cold, rainy northeastern part of the United States In this region and in similar areas such as northern Europe, low temperatures limit the amount of clay that can form. But perhaps more importantly, the high amount of rainfall relative to evapotranspiration (P-E) means that any elements that might have taken part in clay formation are rapidly leached away. Here is how Jenny (1941) described conditions in such cold, humid areas: "A large part of the water added to the soil percolates through the profile and by way of deep seepage and ground water finally reaches the rivers and oceans. Materials dissolved are leached out." Cold, strongly leached environments such as those found in New England and parts of northern Europe allow very little opportunity for clay to form in the soil. Table 4.3 presents temperature and rainfall data from Phoenix, Arizona, and Rangeley, Maine.

Table 4.3. Rainfall and temperature summaries for Phoenix, Arizona, and Rangeley, Maine

City	Total annual rainfall (inches)	Average temperature (°F)
Phoenix	8.0	75
Rangeley	41.7	39

Rangeley receives five times as much rain as Phoenix and is, on average, 36° F colder. It should be no surprise that the soils of the two areas are different. The soils in the Arizona desert are dry most of the year and show little evidence of leaching, and the lack of water causes clay formation to be negligible. The situation in Maine and nearby states is just the opposite. Like the desert, there is little clay formation, but for totally different reasons. The state soils of Maine, New Hampshire and Vermont all belong to the soil order Spodosols, an especially interesting class of soils. Russian peasants and Russian soil scientists referred to them as "podsols" or "ashy soils" because their most striking feature is a prominent gray layer just below the surface that strongly resembles wood ashes. This "ashy" layer is created when iron, aluminum and organic compounds are leached out and deposited at a lower depth, leaving just the gray, uncoated mineral grains behind. A field soil scientist with many years of experience with Spodosols once told me that such soils form in landscapes that "drip water much of the year." C.F. Marbut, one of the founders of soil science in America, described Spodosols (then called podsols) in one of his lectures as "the most thoroughly leached soils of the world" (Jenny 1941).

In addition to Spodosols, The National Cooperative Soil Survey recognizes three other soil orders in Maine — Entisols, Inceptisols and Histosols (Ferwerda et al. 1997). Entisols by definition are soils with little or no evidence of soil development, which effectively rules out much clay formation. Inceptisols (from the Latin *inceptum*, or beginning) are soils with only minimal (or beginning) evidence of soil development, again ruling out significant clay formation. The last of the four soil orders recognized in Maine, Histosols (from the Greek histos, meaning tissue), consist of deep deposits of organic matter, again precluding much clay formation. In summary, very little clay is created in the cold, wet and "dripping" landscapes such as those found in Maine and nearby states.

This chapter has discussed clay formation in four distinct regions of the United States: the deserts, the former grasslands of the nation's interior, the cold, humid Northeastern forests and the warm, humid Southeastern forests. Some might object to this overtly "zonal" approach, arguing that it does not explain everything, and they would be correct. Tying clay formation and other processes to broad climatic/vegetation zones does not capture all of the information one would like, but it captures a lot of it; and in addition, even readers who are new to the study of soil have little trouble understanding the concepts. In that vein, I will introduce yet another region, this time outside the United States.

Lateritic Soils

Keller (1957) described soils of the southeastern United States as being "kaolinitic to lateritic." In using the term lateritic, Keller was referring to the fact that many southeastern soils have a large component of clay minerals that do not contain silica, but instead consist of

Oxides of iron, such as Fe_2O_3 and FeOOH, and

Oxides of aluminum, such as $Al(OH)_3$ and AlOOH.

Iron-bearing clays account for the bright red colors so characteristic of soils in places such as Georgia and Alabama. Clay-sized minerals of iron and aluminum oxides form in warm, high rainfall areas when individual ions of aluminum or iron surround themselves with oxygen ions to form octahedral (eight-sided) structures. The individual octahedral units link together side by side to form sheets; the sheets then combine to form layered structures in which the individual sheets are held together by hydrogen bonds. Some of these oxide clays consist of highly organized crystals, while others have a more indefinite or amorphous form.

Soil layers with very high concentrations of iron and aluminum oxides are commonly called laterites, from the Latin later, meaning brick. This term apparently originated with English geologists who, while working in India during the 1800s, observed the local people cutting soil up into small blocks. After being dried in the sun, the blocks became very hard and were used to build houses and other structures. It was later learned that people in places such as India had been doing this for many centuries. Large deposits of iron and aluminum oxides occur mostly in the tropics, but laterite has been mined in several American states, including Alabama, Georgia and Arkansas. American miners had no interest in using the laterite as bricks, however; instead they were interested in the $Al(OH)_3$ it contained, which also is known as bauxite or aluminum ore. But most bauxite mining today takes place in tropical countries, where large deposits of laterite were formed through the "large removal of silicon and the relative accumulation of aluminum" (Jenny 1941).

Based on Jenny's characterization, one could assume that laterite formation is favored by the leaching of large amounts of silica from the landscape. We will explore that idea by comparing the climates of two areas: the American state of Georgia and the country of Guyana. Guyana is a small country on the northeastern coast of South America, bordering Venezuela to the west and Brazil to the south. Located about 5° north of the equator, it is hot all year long. Below are rainfall and temperature data for Athens, Georgia, and Georgetown, Guyana.

Table 4.4. Rainfall and temperature summaries for Athens, Georgia, and Georgetown, Guyana

City	Total annual rainfall (inches)	Average temperature (°F)
Athens, Georgia	46	63
Georgetown, Guyana	89	80

The average temperature in Georgetown is 17° F hotter than in Athens and Georgetown gets nearly twice as much rain, clear evidence that both weathering and leaching are much more intense there; and the chemistry of river water should reflect this. We test this idea by comparing two rivers, the Chattahoochee in Georgia and the Demerara in Guyana. The Demerara flows east from the Guyana highlands and empties into the Atlantic Ocean at Georgetown. The mining of bauxite or $Al(OH)_3$ is an important part of the nation's economy and much of it is mined in the Demarara River basin. With such extensive formations of bauxite, we would assume that large amounts of SiO_2 are leached from the land and carried away in river water; and the data below, from Clarke's *Data of Geochemistry* (1924), show that this is true.

Table 4.5. SiO_2 content (% of dissolved load) for the Chattahoochee and Demerara Rivers

River	SiO_2 content
Chattahoochee River, Georgia	37.73
Demerara River, Guyana	55.92

Some geologists have described rock weathering as a stepwise separation of elements similar to the process of qualitative analysis in chemical laboratories. If one accepts this analogy and thinks of rock weathering as a global chemical experiment carried out by nature, then bauxite $Al(OH)_3$ could be considered the end point of the experiment. Earlier it was noted that the large amount of silicon dissolved in certain rivers suggests that it is more mobile than iron or aluminum. Field observations such as these, along with laboratory determinations (Correns 1949 and others), suggest that the relative mobility ranking (susceptibility to leaching) of silicon, iron and aluminum is of the order Si > Fe > Al. It is revealing to note that seawater contains about 850 times more silicon than iron and about 3.5 times more iron than aluminum, additional evidence in favor of this ranking. The fact that so much iron is held on the land and so little makes its way to the ocean is of special significance.

"Give me half a tanker of iron and I'll give you an ice age." This statement, attributed to the late John Martin, former director of the Moss Landing Marine Laboratory, is not so farfetched as it might seem. It is based on research showing that many areas of the ocean are "iron deserts" and that sprinkling iron dust in the right place can trigger blooms of plankton covering many square kilometers. Martin and others have argued that creating a number of such "plankton cities" across the ocean might absorb enough heat-trapping carbon dioxide to slow down or stop global warming. Proposals such as this underline the fact that iron levels in the ocean are very low, low enough to limit biological productivity in many areas. The table below shows the average amounts of sodium, iron, and aluminum found in seawater.

Table 4.6. Amount of sodium, iron, and aluminum dissolved in the ocean (parts per billion)

Element	Parts per billion
Sodium	10,800,000.0
Iron	3.4
Aluminum	1.0

You will note the large amount of sodium dissolved in seawater compared with iron. You also will note that seawater contains even less aluminum than iron, but that is a good thing. A lot of aluminum floating around would not be good for life in the oceans, whether plant of animal. But iron is another story. It is an important nutrient, and although widespread seeding of the seas with iron to micromanage global climate might not be feasible, research has shown that adding iron dust to selected areas of the ocean can greatly accelerate plankton growth — and cause a rapid increase in commercial fish populations. In 2012, the Haida Salmon Restoration Corporation dusted an area of ocean in the Gulf of Alaska with 120 tons of iron sulfate. Their objective was to stimulate phytoplankton growth and increase the food supply for baby salmon — and hopefully increase the amount of harvestable fish. It seems to have worked; the following year, the west coast of North America experienced its largest commercial salmon harvest in history (Zubrin 2014). Some are advocating for similar projects in cod fishing areas, but concerned environmentalists are urging caution.

Fertilizing the ocean with iron to increase fish populations or to sequester carbon might or might not become a reality, but the fact that it is being considered highlights the fact that, as the earth has evolved, the oceans have become increasingly salty and increasingly bereft of iron. If you evaporate a cubic foot of seawater, you will be left with about 2.2 pounds (about one kilogram) of salt. The contrast with iron is telling; if you evaporate a ton of seawater, the amount of iron remaining will weigh about as much as a single human eyelash. The eyelash comparison has been attributed to Seth John of the University of South Carolina. If all the salt in all the seas could be removed and spread evenly over the Earth's land surface it would form a layer more than 500 feet or about 150 meters thick, about the height of a 40-floor office building (Kurlansky 2002). If you did the same with all of the iron dissolved in the oceans, the resulting layer would be extremely thin. There is little doubt that low iron levels limit the productivity of both plant and animal life in the ocean.

This chapter has discussed the roles of silicon, aluminum and iron in forming the structure of silicate clays, noting that there are two basic building blocks of silicate clays: (1) the silicon tetrahedron, which consists of four oxygen ions enclosing a single silicon ion; and (2) the aluminum octahedron, which consists of six oxygen ions enclosing a single aluminum ion. The roles of temperature and rainfall in determining both the amount and kinds of clay formed also were stressed. The nature of clay formation in five distinct climatic zones was discussed:

1. Hot deserts of the American Southwest
2. Former grasslands of the American Midwest
3. Warm, rainy forests of the Southeastern United States
4. Cold, rainy forests of the Northeastern United States
5. Hot, very rainy tropical forest of Guyana, South America

The presence of oxide clays (clay-sized minerals of iron and aluminum oxides) in the southeastern United States and in the tropics was noted. Because of the thick layers of bauxite that form in many tropical areas, thick enough to be mined for aluminum ore, some geologists have referred to bauxite or $Al(OH)_3$ as the "end stage of weathering." The immobility of aluminum and the low amounts of it found in ocean waters is of little significance, but iron is another story. There is compelling evidence that low levels of dissolved iron in the world's oceans limits biological productivity.

This chapter has focused on silicon, aluminum and iron — the least mobile of the "big seven" elements that, along with oxygen, make up the near-surface rocks. The next chapter is dedicated to the rest of the big seven — sodium, potassium, magnesium and calcium; and they all have interesting stories to tell. Potassium, for example, plays a surprising role in clay formation and in the soil at large, while many atoms of sodium and calcium take a long journey to the sea — and then they come back. How these two elements return to the land, the different journeys they take, and the importance of these journeys to terrestrial life make for an interesting narrative.

CHAPTER 5. TO THE SEA AND BACK

The last chapter focused on silicon, iron, and aluminum, the least mobile of the rock-forming elements. It was shown that large quantities of them remain on the land to form soil clays. This chapter focuses on the four remaining elements that dominate the structure of surface rocks; sodium, calcium, magnesium, and potassium, and what happens to each of them after weathering agents remove them from rocks. The table below shows the relative susceptibility of each to being leached from the land. Why do they differ? For example, why does potassium tend to stay on the land, while sodium ions are carried away to the sea in vast quantities?

Table 5.1. Relative mobility of sodium, calcium, magnesium, and potassium

Element	Relative mobility
Sodium	100
Calcium	45
Magnesium	30
Potassium	20

Calcium and Magnesium

As discussed earlier, one of the reasons magnesium is less mobile than calcium is that magnesium commonly finds its way into the structure of 2:1 clays such as montmorillonite, and large amounts of such clays are found in soils, especially in arid and semiarid climates. The Mg^{2+} ion, with a radius of 0.66 Å, is similar in size to the Al^{3+} ion (0.51 Å), so magnesium readily substitutes for aluminum in 2:1 clays such as montmorillonite. Magnesium also is a component of a clay mineral called chlorite. According to Grim (1953), chlorite consists of "alternate mica-like and brucite-like layers." The brucite-like layer to which Grim refers has the general composition of $Mg(OH)_2$, which shows why the mineral chlorite also serves as a depot for magnesium in the soil. These two minerals, montmorillonite and chlorite, are two reasons that more calcium than magnesium is leached from the land and carried away to the sea.

Potassium

But potassium is more of a puzzle. A very large ion, with a radius of 1.32 Å, it cannot fit into the center of any octahedron or tetrahedron, which are the building blocks of clay, so why is it so much less mobile than calcium, magnesium, or sodium? What is holding it on the land? The answer, or at least a big part of it, can be seen in a host of soils, including many of those in central North Carolina where I live. Many of the soils here weathered from granite or similar igneous and metamorphic rocks, and it is possible to discern the character of the parent rock by examining a sample of the red clay subsoil common to the area. By moistening a small sample and squeezing it out into a thin ribbon, you can see or feel remnants of the quartz, feldspar, and mica that made up the original rock. If you rub a moist sample between your thumb and fingers, you can feel the large, angular sand grains that are the remains of quartz, while feldspar is the main source of the sticky clay that coats your fingers and colors them red. A third major component, and the one which is of most interest here, can be seen as the little flakes of mica that glitter and twinkle with reflected sunlight (a 10X magnifying glass is helpful). Mica, a component of granite and many other kinds of rock, is where much potassium is sequestered on land.

Micas

Mica is highly distinctive in appearance because large chunks of it easily pull apart into thin, often flexible sheets. In general, micas are divided into two groups: those that are light colored such as muscovite and those that are dark colored such as biotite. In medieval times, the English referred to muscovite as Muscovy glass because of its widespread use in Czarist Russia as a substitute for window glass. Biotite is named in honor of a French physicist whose last name was Biot. The chemical formulas for the two main types of mica are informative, although a bit complicated.

Biotite: $K (Mg,Fe)_3 AlSi_3O_{10} (OH)_2$

Muscovite: $K Al_2 (AlSi_3O_{10}) (F,OH)_2$

These rather intimidating formulas are presented for two reasons. One is to show that both biotite and muscovite contain potassium (K). The other reason is to show that biotite contains iron, while muscovite does not. Biotite is sometimes referred to as "black mica" or "iron mica." Partly because of its iron content, biotite weathers rapidly; as a result, it is not found in soils nearly so often as muscovite (remember how readily iron reacts with oxygen and what that does to mineral structures). Also note that both biotite and muscovite contain potassium, which is the elemental "glue" that holds the individual mica sheets together. Below are excerpts from the official description of the Fannin soil series at its type location in Alleghany County, North Carolina.

1 to 4 inches (2.5 to 10 cm); brown loam; common flakes of mica

4 to 8 inches (10 to 20 cm); brown loam; common flakes of mica

8 to 14 inches (20 to 35 cm); yellowish red clay loam; common flakes of mica

14 to 26 inches (35 to 66 cm); red clay loam; many flakes of mica

26 to 33 inches (66 to 84 cm); yellowish red loam; many flakes of mica

This description continues to a depth of more than 60 inches (1.5 meters), with the content of mica increasing with depth. Note the phrase "flakes of mica," reflecting the sheet structures from which they weathered. All or nearly all of the mica in this soil was inherited from the parent rock, but in 1937 mineralogists identified some clay-sized mica in Illinois that had not come from parent rock but instead had formed in place. This clay-sized mica, which was named illite, is now known to be relatively common in soils. Mica serves as a storage depot for potassium in the soil; and that is a good thing, because only a small amount of potassium is stored on clay surfaces at any given time. Below are some typical data showing the percentage of negative charges on clay surfaces occupied by different elements.

Table 5.2. Percentage of clay charges occupied by Ca, Mg, K, or Na

Data from Black (1957)

Type of soil	Ca	Mg	K	Na
Acidic (n = 9)	69.4	25.1	2.8	2.7
Neutral to alkaline (n = 6)	62.1	25.6	3.9	8.4

Calcium nearly always dominates the exchange complex of clays, while the amount of potassium held is much smaller. Since plants take in large amounts of potassium, it is fortunate that many soils are able to replace the rather small stock of exchangeable potassium whenever it becomes depleted. The replenishment of potassium from micas in the soil is an important and well-recognized phenomenon. Bray and DeTurk first suggested in 1939 that the potassium in mica is slowly released into the soil to be taken up by plants and that this stored potassium is in equilibrium with that on the surfaces of clays and organic matter.

The name potassium is derived from "potash," referring to wood ashes. People learned centuries ago that they could concentrate potassium salts by burning wood and then mixing the ashes with lye and boiling the mixture in a pot, hence "potash." The soluble potassium that resulted was a valuable industrial commodity, useful for making such things as glass, soap, and gunpowder and for dyeing and bleaching fabrics. Prior to about 1860, a lot of hardwood trees in America were cut and burned solely for the purpose of making potash. The first United States patent, issued in 1790 and signed by George Washington, was for an improvement "in the making of Potash." With so much potassium sequestered in the structure of mica and in plant tissue, it is no surprise that comparatively little (compared with calcium and sodium) is leached from the land and carried away to the sea.

We have now established that magnesium and potassium, in addition to silicon, iron, and aluminum, are involved in the formation of soil clays. This leaves calcium and sodium, both of which are simply the wrong size to fit well in either octahedral or tetrahedral structures. As a result, large amounts of both, but more sodium than calcium, are leached from the land and carried away to the sea. A glance at Table 5.1 provides a hint as to why sodium is more mobile than calcium. Very simply, calcium (Ca^{2+}) ions dominate the exchange surfaces of soil clays, so much more calcium than sodium is held in the soil at any given time. In addition, living plants (and animals, especially those with bones) take up and store much more calcium than sodium in their tissues. Nevertheless, a lot of calcium is leached from the land along with sodium and carried away to the oceans. Fortunately, nature has devised a way for some of the calcium and sodium leached from terrestrial landscapes to be returned to the land. The two ions travel to the sea in the same way and often together, but their return journeys can be very different.

The Carbonate Ion

The fate of calcium ions (Ca^{2+}) is closely linked with that of the carbonate ion (CO_3^{2-}). The CO_3^{2-} ion consists of a central carbon ion (C_4^+) surrounded by three oxygen ions in a roughly planar structure. The very small C_4^+ ion (0.15 Å) fits snugly in the hole created by the surrounding oxygen ions, making for a very strong structure, one that can survive the often long and perilous journey to the sea. In river water, the HCO_3^- ion is the most common form. For every dissolved ion with a positive charge such as Na^+ being carried down a river, there must be a balancing ion with a negative charge, and bicarbonate (HCO_3^-) usually serves that purpose. As a result, the world's rivers carry large amounts of dissolved HCO_3^- ions (typically listed as CO_3^{2-} in data tables). On average, the bicarbonate ion accounts for about 35 percent of the dissolved matter in the world's rivers.

Rock weathering never ceases, takes place over vast areas, and has been going on for a very long time. As a result, a lot of carbon dioxide has been washed from the atmosphere, converted into CO_3^{2-}, and then carried away from the land. In oceans, especially the warmer ones, a host of organisms make good use of this CO_3^{2-}, extracting it from the water and combining it with Ca^{2+} to form protective exoskeletons of $CaCO_3$. Eventually the dead bodies of these organisms sink to the bottom where, over a period of many centuries, their skeletal remains form thick layers of limestone. But this limestone does not always stay submerged. With the passing of time, much of it returns to the land surface. This can happen for a number of reasons, such as inland seas drying up, ocean levels dropping, land surfaces rising, or colliding continents lifting former sea floors onto the tops of mountains. One can only marvel that limestone deposits, some of which lay beneath the sea for millions of years, now make up nearly one fifth of the earth's dry land surface.

Soils readily form from limestone, or rather from the sand, silt, and clay impurities that are left behind after the limestone dissolves. Soils can form in nearly pure limestone (≤ 10 percent impurities) or in "limey" materials such as calcareous shale or calcareous sandstone (≥ 50 percent impurities). In any case, soils develop from the mineral particles left behind after the limestone is dissolved, not from the limestone itself. It is easy to imagine why the purity of limestone varies so widely. In some marine environments, such as quieter, deeper waters farther from shore, relatively pure limestone can form because sediments washing from the land settle out closer in. At the same time, in the turbulent near-shore waters, a lot of sand, silt, and clay are mixed with the shells and coral that fall to the bottom, resulting in a "dirtier" limestone. Large deposits of very clean limestone also formed as ancient coral reefs that later were uplifted to become part of the land surface.

Pure limestone is similar to salt (NaCl) in structure and dissolves in much the same way, but it is much less soluble; a gallon of water at pH 7.0 will dissolve only the amount of $CaCO_3$ equal to the size of a pinhead. But water with carbon dioxide dissolved in it is another story. When CO_2 dissolves in water, hydrogen ions are created, and these tiny, aggressive protons immediately begin to attack and break down limestone. Here is the overall equation for the breakdown of limestone by water with dissolved carbon dioxide (a weak solution of carbonic acid): $CaCO_3 + CO_2 + H_2O \Rightarrow$ Soluble $Ca(HCO_3)_2$. The limestone is simply dissolved and carried away, while impurities such as sand, silt, clay, and organic matter are left behind to become the starting material for soil. River water reflects this kind of weathering. For example, Ca^{2+} and HCO_3^- ions together account for 70 percent of the dissolved load in the Shenandoah River, which drains an area of limestone in West Virginia (Clarke 1924).

Sodium and Rain

Much of the calcium that is returned to the land arrives by way of tectonic events such as colliding continents and rising mountains. The return of sodium is less dramatic but no less interesting; it is intimately related to rain and the way in which it forms. People have wondered about the origin and nature of rain for thousands of years. Lucretius wrote more than 2,000 years ago, "Attend now, and I will explain how rain collects in the clouds above, and how the showers are precipitated and descend upon the earth." He went on to say that rain was caused by wind pressing against swollen clouds and that thunder and lightning came from the sparks and noise created when clouds collided, not unlike what happens when two stones are struck together. We should not be too critical of Lucretius. Considering the information available to him, his musings were not so unreasonable.

By 1900, scientists had learned more about clouds and rain than Lucretius could have imagined, but the true origin of the phenomenon remained a mystery until the last half of the 20th century. Early work pointing the way came from French scientist Paul Jean Coulier, who in 1875 began studying the behavior of fog in an enclosed chamber. To his surprise, he found that after a number of experiments, he could no longer get fog to appear unless he let in some fresh air from the outside. He concluded that in order for fog to form, moisture had to condense around the tiny, mostly invisible dust particles that were always present in the air. John Aitken, a Scottish physicist and meteorologist, came to a similar conclusion, writing in 1881,

> Everything in nature which tends to break up matter into minute parts will contribute its share. In all probability the spray from the ocean, after it is dried and nothing but a fine salt-dust left, is perhaps one of the most important sources. From this it will be observed that it is not the visible dust motes seen in the air that form the nuclei of fog and cloud particles. The fog and cloud nuclei are a much finer form of dust, are quite invisible, and though ever present in enormous quantities in our atmosphere, their effects are almost unobserved.

Shortly before World War II, some American scientists began to take a close look at the particle or "salt theory" of rain formation. Alfred Woodcock, at the Woods Hole Oceanographic Institute in Massachusetts, carried out some of the earliest and most useful research. Scientists had known for many years that in order for a raindrop to form, millions of tiny cloud droplets had to come together. Cloud droplets are so small that their dimensions are commonly measured in microns (μ), which is about 1/100th the thickness of a human hair. When a particle reaches 50 μ in diameter, it becomes just visible to the human eye. Cloud droplets have a diameter of about 20 μ (about 1/5 the diameter of a human hair). Let us compare that with an average-size raindrop, which is about 2,000 μ (or 2 mm) in diameter. As made clear by Blanchard (1966), this size difference is a problem.

> It takes about one million cloud drops to make a single raindrop. And there, quite simply, is the problem that faced the atmospheric scientist... How is it possible for such a fantastically large number of cloud drops to get together to form each of the millions of raindrops falling from a cloud when it rains?

Scientists agreed that tiny cloud droplets must come together in order for raindrops to form, but they could not figure out exactly how it happened. Some speculated that the process might be similar to what happens when a snowball is rolled down a steep hill. As the snowball rolls, additional snow sticks to the surface, causing it to become larger and larger. If a cloud drop could be induced to start falling, researchers reasoned, it would behave much like an airborne snowball, picking up droplets as it fell and eventually becoming a full-fledged raindrop. They knew that the beginning cloud droplet would have to be much larger than normal in order to start the process, but exactly how big?

In 1946, Langmuir and Blodgett calculated what they called "collision efficiencies," demonstrating that cloud droplets with a radius of 20 μ or larger (about one fifth the diameter of a human hair) would be big enough to pick up additional water and form raindrops as they fell from the sky. But in order to form one large snowball droplet, at least eight regular-size cloud droplets would have to combine. Getting eight individuals to come together in pursuit of a common goal, even if the individuals are cloud droplets, usually requires some kind of inducement; and as mentioned previously, it had long been speculated that salt particles might be a big part of the answer—forming the nuclei around which raindrops form.

But after scouring the scientific literature, Woodcock could find little information about how many salt particles might exist in the atmosphere at any given time or how large they might be. Very few measurements had been taken and it seemed that none had been made at elevations where clouds typically form. Woodcock decided to remedy that and came up with a way to do it. He held a small, carefully cleaned glass plate out the window of a small, low-flying airplane for a prescribed number of seconds, allowing salt particles in the air to strike the plate and stick to its surface. He then took the plate back to the laboratory and carefully counted the salt particles. After much experimentation, he learned how to compute the number of salt particles in a given volume of air. He found that there were huge numbers of salt particles above the sea surface and that winds often carried them thousands of feet into the air. He also determined the size distribution of cloud droplets, finding that many of them were large enough to form raindrops. The last piece of the puzzle had fallen into place. As Blanchard (1966) later wrote, Woodcock's measurements showed that

> The atmosphere is like a giant salt shaker which the sea continually is filling with salt particles. Then, in turn, the atmosphere pours the sea-salt into the clouds. The largest of the salt particles provide

the nuclei for raindrops. They are the snowballs that you start rolling down the hill.

But a problem still remained: How does so much salt, often nearly a million particles in every cubic meter of air, get from the sea into the air? Simple evaporation of saltwater would not do it; the water would leave the surface, but the salt would be left behind. A clue as to what was going on came from an unlikely source. In the late 1920s and early 1930s, engineers and physicists were trying to explain the origin of the tiny water droplets commonly observed in the air above the water surface in steam boilers. Otto Stuhlman (1932), a biophysicist at the University of North Carolina, came up with the most probable answer. He had observed that bubbles constantly formed, burst, and then reformed on the surface of the water in boilers. Stuhlman reasoned that tiny water droplets must be launched into the air by the pencil-like jets of water he could imagine rising from these bubbles as they burst.

In 1937, W.C. Jacobs was studying the distribution of salt in the air above oceans. After reading Stuhlman's work, he surmised that salt particles in the air above oceans might originate in the same way. The start of World War II interrupted research in this area, but in 1950 Alfred Woodcock, looking for a way to explain the origin of all the salt he was collecting on his glass plates, began to examine Jacob's sea-bubble hypothesis. The problem was that no one had ever seen these bursting bubbles and the tiny pencil-like jets that supposedly launched water droplets (with their salt) into the air. Stuhlman, Jacobs, and others had only imagined them.

Woodcock and his colleagues at Woods Hole decided to remedy that. They spent many hours in the laboratory peering through microscopes at bursting bubbles, but were disappointed; whatever happened took place too rapidly to be observed by the human eye, so they tried another approach. In 1953, they obtained a high-speed movie camera, and Charles Kientzler, after solving some daunting technical problems, spent an entire summer taking movies of bursting bubbles. Blanchard (1966) wrote,

> By summer's end Kientzler's pictures had removed any doubt about the existence of a bubble jet... When a bubble broke and was in the process of collapsing, a vertical column of water...looking for all the world like a miniature Eiffel Tower, rose swiftly from the bottom of the collapsing bubble... A bubble jet usually produces about five drops, one after the other, and all extremely rapidly.

You might assume that bubbles form only at the beach as wave after wave strikes the shoreline and the surf becomes a seething, frothy mass;

but bubbles constantly form and burst all over the open ocean. When the wind blowing across the surface exceeds a speed of about six or seven miles per hour, whitecaps are created. With increasing wind speed, the waves become unstable; they then curl over and break apart, or their tops are blown off by the wind. As the wave tops crash back into the sea, they carry globs of air into the water, but the air quickly comes to the surface again as trillions of small bubbles. Little "Eiffel Towers" of water jet up from the bursting bubbles, dribbling little drops of salt water into the air. The wind carries these jet drops high into the atmosphere where they dry out and become the sea-salt particles that are so instrumental in forming rain. Since oceans make up so much of the earth's surface, the lower atmosphere (where rain is made) should contain a lot of salt particles, and it does. Sea salt accounts for nearly 80 percent of the particulate matter in our atmosphere (Table 5.3).

Table 5.3. Sources and amounts of particles in the atmosphere

Data from Allen (1983)

Source	Millions of tons/year	Percent of total
Sea salt	1,000	77.0
Soil dust	200	15.4
Volcanoes	4	0.3
Forest fires	3	0.2
Human activity	92	7.1
Total	1,299	100.0

If, as shown in the table, sea salt accounts for 77 percent of the particles in the air, then rain should be slightly salty, and studies have verified that it is. From August 1962 to July 1963, the United States Geological Survey studied the chemical composition of rainfall in a 34 square mile (88 square kilometer) area in southeastern Virginia and eastern North Carolina. They measured the total amount of rainfall each month and determined the amounts of different ions dissolved in the rainwater. During the twelve-month period from August 1962 to July 1963, rain deposited an average of 13.5 pounds (six kilograms) of salt (NaCl) on every acre of the study area (Gambell and Fisher 1966).

Ocean bubbles generate more than 70 percent of the rain-forming particles in the atmosphere, but oceans also make up about 70 percent of the planet's surface, so most of this rain falls back into the ocean and is of no benefit to the land. Which brings up an interesting question: How much of the rain that falls on land originates from salt particles and how much from soil dust or from particles created by volcanoes, forest fires, or industrial activity? Junge and Gustafson (1957) measured the average amount of chlorine, an indicator of the amount of salt, in the rainfall at 62 weather stations across the United States. They found that "The concentration decreases rapidly with increasing distance from the coast, tending toward a constant level inland." Measuring northward from Mobile, Alabama, and northwestward from Cape Hatteras, North Carolina, the rapid decrease in concentration ceased at about 500 kilometers (310 miles) from the coast and then remained relatively uniform throughout the interior of the country.

They concluded that this rapid depletion with increasing distance from the ocean was partly the result of salt particles being washed out of the lower atmosphere by falling rain. The other reason for the apparent drop off was that with increasing distance inland, salt particles are increasingly carried up to higher altitudes and mixed throughout the troposphere—a dilution effect. Because of this, we would expect salt particles to play a greater role in rain formation in coastal areas than in continental interiors, and they do. Junge (1958) determined the average ion concentration in rainwater over the United States (June 1955 through July 1956) for weather stations located throughout the country. Using these data, I compared rainfall composition for ten stations located in the center of the country with that of ten stations located along the Atlantic coast east of the Appalachians. Below are summary data for calcium, sodium, and what meteorologists call "excess sodium." "Excess sodium" is the sodium in rain that is not matched by an equivalent amount of chloride. The ratio of Cl⁻ to Na⁺ in seawater is 1.800 (Sverdrup et al. 1946). If this ratio is found in rainfall, one can assume that all of the raindrops coalesced around salt particles. However, that almost never happens and the amount of variation from this ratio can be used to estimate the relative proportion of the rain that originated from sea salt and soil dust respectively. Listed below are summary data from the midcontinent and from the Atlantic coast. Note: The calculations in Table 5.4 assume an annual rainfall amount of 26 inches (66 centimeters) in the midcontinent (average for Des Moines, Iowa) and 46 inches (117 centimeters) for the southeastern Atlantic coast (average for Athens, Georgia).

Table 5.4. Amount of sodium, excess sodium, and calcium deposited by rain and snow (pounds per acre) annually

Calculated from Junge (1958)

Location	Na^+	Excess Na	Ca^{2+}
Midcontinent	2.2	82 percent	16.2
Atlantic coast	3.7	49 percent	3.7

This table illustrates several important points. First, precipitation nearer the coast carried much more sodium to the ground than it did in the continental interior. Second, more than 80 percent of the sodium in the continental interior was "excess," compared with about 50 percent along the Atlantic coast. This suggests that most of the sodium reaching the ground in places such as Iowa and Missouri came from soil dust or other particles that originated on land. However, sea salt particles do make their way to the interior of the continent and play a role in rain formation even there. Sea salt plays a bigger role along the coast, but even there it accounted for only about half of the nuclei that formed raindrops. Soil dust, volcanoes, forest fires, and human activity account for less than one fourth of the total particulate matter in the earth's atmosphere. However, they probably initiate well over half of the total precipitation on continental landmasses.

The last point to be taken from Table 5.3 is that very large amounts of calcium were deposited by precipitation in the center of the country compared with the rather low amount along the Atlantic coast. Keller described this process of nutrient enrichment in 1957, stating that landscapes in the Midwest farm belt "are replenished continually by nutrient reserves contained in rock and mineral particles blown onto the soil by the wind from the Missouri River flood deposits and from the western dry plains."

The Salt Lick Problem

Although every rainstorm brings sodium to the ground, helping to replace that lost to runoff and leaching, there is still a problem, or at least many people think there is; and it goes something like this: Most plants do not need sodium, so they take up very little of the element and store very little of it in their tissues, and this raises a question. If plants contain so little sodium, how do animals that eat nothing but plants get the salt they need to stay alive? Many people think that salt licks are the answer. Mark

Kurlansky published a popular book in 2002 titled *Salt: A World History*. Although focusing mostly on the history and economics of salt, Kurlansky ventured briefly into natural history by asserting that because grazing animals cannot obtain enough salt from the vegetation they eat, they must get it by "finding brine springs, salty water, rock salt, any natural salt available for licking." He went on to say, referring to North America, that salt licks are found "throughout the continent." Kurlansky is not alone in making this assertion; it is repeated again and again in scientific papers and in popular articles. Here is an example from the scientific literature (Subbarao et al. 2003).

> Because of the contrast between plants and animals in their electrolyte requirements, there is insufficient Na available in the edible portions of most plants for large herbivores. Therefore, the dietary requirements of herbivores for Na must be met from external supplements, such as salt licks.

There is an element of truth to such statements. Salt licks do occur in some parts of the country, notably in areas underlain by layers of sedimentary rock. Famous salt licks have been described, for example, in Kentucky, Ohio, Indiana, and West Virginia. But there is a problem; there are many millions of acres on which salt licks are extremely rare or nonexistent. This includes large parts of eastern states such as North Carolina, South Carolina, Alabama, and Georgia. Due to the nature of the underlying rocks and the fact that there is such abundant rainfall evenly distributed throughout the year, even if salt were exposed at the surface it would quickly be leached away.

I have questioned foresters, soil scientists, and others who worked in the southeastern United States for many years; and despite having walked over thousands and, in some cases, hundreds of thousands of acres, not one had ever to his or her knowledge seen a natural salt lick. If natural salt licks exist in the area, they must be very rare. Some wild animals in certain locations do have access to salt licks, but in many parts of the country a deer or antelope might live out its life without ever encountering one. Kennedy (2006), writing in the North Carolina Encyclopedia, offers an interesting local perspective.

> American Indians, and later European settlers, held salt in high esteem. Since the region [North Carolina] lacked natural salt, even the salt "waste" from preserving hides and meats was put to use. The Indians carefully saved the salt remnants and placed them strategically on their hunting grounds to attract wild animals and birds. These recycled salt licks were enhanced at every opportunity with additional

salt scraps. They became tribal property, not to be infringed upon by neighbors, although they were often discovered and seized by settlers. Local names such as "Big Lick," "Licking Creek," and "White Lick" appear in various parts of the state. These salted areas lasted for centuries, perhaps with some assistance by later settlers who dumped their salt scraps on the old salt licks.

Even the most ardent believers in the salt lick theory would have to concede that only some grazing animals, and probably a small minority, have access to non-plant sources of salt; most animals are forced to live out their lives in areas where saltlicks simply do not exist. How do such unlucky creatures manage to survive and reproduce without the supplemental salt that they supposedly require? The answer is that grazing animals can get all the salt they need just from eating grass. Many have mistakenly assumed that grasses are low in salt content because most of the plant foods that humans eat, such as fruits, nuts, and vegetables, are very low in sodium. But that assumption is incorrect, as the following analysis will show.

Ideally, according to most health authorities, an adult human should consume about 500 to 1,000 milligrams of sodium daily. Table 5.5 shows the quantities of various foods, such as meat, nuts, fruits, grains, and nonleafy vegetables one would have to consume daily to get that much. The data are from the USDA National Nutrient Database for Standard Reference, which contains nutrient information on about 9,000 foods. The value for each food group in Table 5.5 is the average of ten individual foods. For example, to obtain the overall value for game meat, sodium contents were averaged for antelope, bison, bear, wild boar, deer, horse, muskrat, opossum, rabbit, and squirrel. In the case of grains, individual sodium contents were averaged for wheat, oats, rice, barley, buckwheat, millet, corn, quinoa, sorghum, and spelt. The members of the other food groups were similarly diverse.

It is clear from Table 5.5 that early humans would have had trouble getting enough salt if they ate only plants. There is no way a human being could eat more than 50 pounds of legumes per day or more than 220 pounds of fruit; but meat is a different story. Using calorie information from the USDA nutrient database, it was determined that if an early human ate only meat and consumed 3,000 calories per day (about 3.5 pounds), he or she would be taking in about 1,100 milligrams of sodium daily. But an all-meat diet was unlikely in most environments, so most hunter-gatherers undoubtedly ate a quantity of foods very low in sodium (such as nuts, grains, and vegetables) in addition to meat or fish. This means that the most likely range of daily sodium consumption for early humans was perhaps 500 to 1,000 milligrams per day, well within the healthy range.

Table 5.5. Weight of different foods yielding 1,000 milligrams of sodium

Type of food	Weight (pounds and kilograms)
Game meat	3.2 (1.4 kg)
Fish	2.9 (1.3 kg)
Vegetables	34.4 (15.6 kg)
Grains	49.0 (22.2 kg)
Legumes	51.2 (23.2 kg)
Nuts	88.1 (40.0 kg)
Fruits	220.3 (100.0 kg)

Omnivores (such as humans) and carnivores (such as lions, wolves, and tigers) could get sufficient salt by eating animal flesh, but what about grazing animals? How did antelopes, buffalo, horses, and aurochs (the ancestors of modern cattle) manage to get enough salt if they ate only plant tissue? The answer is that, unlike humans, grazing animals do not rely on fruits, vegetables, or nuts for food; instead they eat mostly grass. The common wisdom is that grazing animals get only small amounts of sodium from the grass they eat and must supplement it by regularly visiting salt licks. But those who assert this have never really evaluated the sodium-supplying potential of grass; they simply have assumed that since other plant foods are low in sodium, grasses must also be low in sodium. But, as we will demonstrate, that is a false assumption.

Smith et al. (1978) studied the sodium content of pasture grasses in New Zealand by analyzing data from 2,767 foliage samples from all parts of the nation. This very large database provides insight into the sodium-supplying potential of grasses. The average sodium content of the 2,767 New Zealand samples on a dry-weight basis was 0.20 percent. Starting with this number, which is fairly typical for grasses, we can show that grazing animals, with certain unusual exceptions such as dairy cattle, do not need supplemental sources of sodium. We do this by comparing the sodium intake of a man eating only meat with that of a horse eating only grass. There still are a lot of horses in the world (more than nine million in the United States alone), and a lot is known about their nutritional requirements. According to the experts, a 1,000-pound (454 kilogram) horse needs to consume about 20 to

25 pounds (nine to eleven kilograms) of forage daily on a dry-weight basis. In contrast, a man eating only meat needs to consume about 3.5 pounds (1.6 kilograms) daily to acquire 3,000 calories.

Let us first consider the sodium content of meat. According to the USDA nutrient database, our ten species of game animals averaged 70 milligrams of sodium for every 100 grams of flesh; but this is based on the "wet weight" of meat. On average, meat contains over 60 percent water, so 100 grams of meat is only about 40 grams on a dry-weight basis. After subtracting out the water content, the average sodium content of our ten game species (percent of dry weight) was calculated as follows:

There are about 70 milligrams of Na in 40 grams (dry weight) of meat.

40 grams = 40,000 milligrams

70 mg of Na / 40,000 mg of meat (70 / 40,000) = 0.18 percent Na.

These numbers go against the conventional wisdom, but they provide an accurate picture. Grasses, the typical food of grazing animals, contain as much sodium (and thus salt) on a dry-weight basis as does meat (0.20 percent versus 0.18 percent). Let us take this comparison further by assuming that an adult human male weighing 150 pounds eats nothing but meat on a given day. To get 3,000 calories, he would have to take in about 3.5 pounds of cooked meat. Since cooked meat, on average, still contains about 60 percent water, this would be about 1.4 pounds (0.6 kilogram) on a dry-weight basis. This amount of meat would provide him with about 1,145 milligrams of sodium, or about 7.6 milligrams of Na per pound of body weight (1,145 milligrams / 150 pounds (man's weight) = 7.6 milligrams Na per pound of body weight.)

Let us now compare the amount of sodium a human gets per pound of body weight from eating meat to the amount a horse (or other grazing animal) gets from eating forage. Assume a horse eats 20 pounds or 9,080 grams (20 pounds x 454 grams/pound) of forage per day (dry weight) and that the forage contains 0.20 percent (0.0020) Na as shown in the massive New Zealand database.

9,080 grams of forage daily x 0.0020 = 18.2 grams Na per day

18.2 grams = 18,200 mg

18,200 milligrams Na /1,000 pounds (body weight of horse) = 18.2 milligrams of Na per pound of body weight

The calculations above are a little tedious, but it was important to go through them because the end result is so contrary to what has been so widely believed for so many years. When you do the calculations, it appears that grass, the food of grazing animals, supplies as much sodium, if not more, than the meat eaten by carnivores and omnivores. In the research cited here, the sodium content of grass was 0.20 percent on a dry-weight basis, compared with 0.18 percent in the meat of game animals. Even assuming that the sodium content of grass and meat is the same, the grazing animals still have an advantage. Their plant diet is much lower in energy density, so they consume a higher percentage of their body weight in food each day, taking in more sodium in the process. In our example, the grass-eating horse consumed 18.2 milligrams of sodium for each pound of body weight. This is more than twice as much as the meat-eating human, at 7.6 milligrams per pound of body weight. Table 5.6 summarizes the average sodium content of the major plant food groups, adding grass to the list.

Table 5.6. Sodium content of different food groups (percent of dry weight)

Food Group	Percent sodium
Grass	0.200
Meat	0.180
Vegetables	0.070
Legumes	0.010
Fruits	0.006
Grains	0.005
Nuts	0.003

The near equivalency of grass and meat lends credence to the oft-quoted aphorism, "All flesh is grass." That is why the salt lick narrative never rang true. It is hard to imagine that nature would have allowed such a crucial link in the ecosystem (i.e., the availability of salt, which all animals must have) to be held hostage to random geological accidents such as the presence of salt licks or salty springs. As shown in Table 5.5, grass contains as much sodium as meat and many times as much as any other plant food. There is a reason for this. When you eat most plant foods, such as an apple, a nut, a tomato, a bean, or a cucumber, you are eating something that the plant created for a specific purpose—most often reproduction. Since sodium does not play a role in the growth and reproduction of most plants, little of it is found in such structures. But much of the tissue in the roots, stems, and leaves of grass is devoted to the uptake and translocation of soil water—so an animal eating grass stems and leaves (especially stems) is by default consuming water that recently was extracted from the soil and nearly all soil water contains salt. Grazing animals, because they specialize in eating the water-conducting tissues of plants, have primary access to the sodium extracted from the soil. When they eat stems and leaves, they are taking in slightly salty soil water, water that came to earth as slightly salty rain only a short time before.

The Celery Problem

Having addressed the salt lick issue, let us now consider another salt-related issue, one that is more closely related to humans. It has to do with a vegetable, or more precisely, a class of vegetables and why we eat them. You might have read or heard that celery has negative calories, meaning that you expend more energy chewing and digesting it than your body absorbs after eating it. This might or might not be true, but if one takes the energy costs of growing, harvesting, transporting — and chewing into account, celery as a source of food energy is a net loss. The same would be true of a number of other vegetables such as radishes, lettuce, spinach and watercress. Despite this, it is easy to see why modern people eat such foods; they provide variety to the diet and contain important vitamins and minerals.

But what about early humans, who were mostly concerned with acquiring calories and knew nothing of vitamins and minerals? Since we now grow and eat such foods, it is fairly certain that our ancestors ate the ancestors of plants such as celery and radishes, but why would they have done so? If one takes the energy costs of seeking out such foods and harvesting them into account, the net energy gain would have been minimal or nonexistent. Time would have been much better spent in other ways. On any given day, our foraging ancestors could spend their time in more calorie-yielding pursuits, such as picking berries, hunting wild pigs, gathering shellfish, harvesting nuts, or pilfering eggs from nesting birds.

Like other omnivores, foraging humans had a problem to solve. Of the 50 or so foods that could be eaten in an area, what were the ten, 15 or perhaps 20 that would yield the most food energy for the least expenditure of time and effort? Fortunately, most early human groups (those that survived and passed on genes) managed to work this out through trial and error; they probably required only a generation or so in a new area to arrive at what some ecologists call an "optimal foraging strategy" (Emlen 1966, MacArthur and Pianka 1966). Humans learned very early to concentrate their efforts on those foods that were the most plentiful, the easiest to acquire and the most energy rich, often seeking out those that were sweeter or had more fat. In the wild, such foods usually were the most health promoting, since they provided the most protein, essential fats, vitamins, and minerals.

So why would early humans have spent their time and energy gathering the ancestors of celery, carrots, spinach or radishes? It is likely that both grazing animals and early humans sought out and consumed such foods because of their high salt content. We have shown that, on a dry-weight basis, game meat, fish, and grass contain about 0.20 percent Na. This equates to a salt (NaCl) content of about 0.5 percent. The table below shows how this compares with the amount of salt in celery, radishes, and other foods we commonly refer to as salad vegetables.

Table 5.7. Salt (NaCl) content of selected foods

(Data from USDA Nutrient database)

Food	NaCl content (% dry weight)
Game meat	0.50
Fish	0.50
Grass	0.50
Beets	1.60
Carrots	1.50
Turnips	2.08
Radishes	2.08
Watercress	2.11
Spinach	2.31
Celery	4.41

The average salt content of game meat, grass and fish is 0.50 percent (dry weight basis). In comparison, the average salt content of commonly eaten salad vegetable in the list above is 2.30 percent, more than 4.5 times greater. Grazing animals acquired life-giving salt from the grasses they ate and then supplied this salt to human hunters who consumed them. Although early humans could not acquire salt directly from grass as grazers did, they were able to eat the conducting tissue (roots, stems and leaves) of succulent, salt-rich forbs that grew along with the grasses, harvesting some of the salt that had been carried down by rain and stored temporarily in the soil water.

Changing Direction

Thus far we have migrated through a number of scientific disciplines. In Chapter 2, which took us from the Big Bang to the formation of rocks at the earth's surface, we operated in the realms of cosmology, geology, and physics. Chapter 3, which focused on the structure of rocks and the weathering agents that break them down, drew heavily from the basic principles of chemistry. This chapter dealt with the nature of rain, which brought us into the domain of meteorology; and finally, in examining the idea that grazing animals need salt licks to survive, we utilized data and concepts from agronomy and animal husbandry. Despite ranging so widely across disciplines, our discussion has had a fairly narrow perspective, focusing largely on the mineral world—with rock formation, rock weathering, and the fate of mineral elements released by weathering.

Beginning with the next chapter, our focus narrows in one sense while it broadens out in another. It narrows by focusing more on the soil itself and on a particular soil horizon: the surface layer, or what is commonly referred to as the A horizon. This dark, seemingly fragile layer, which blankets nearly every bit of land on earth that supports vegetation, is softer, looser, and much higher in organic matter than the soil layers beneath. Otherwise, at least to the casual observer, there is nothing really remarkable about it; just the uppermost part of the mineral soil darkened by organic matter. But if all biological activity in soil A horizons all over the world were to cease, many species on Earth would become extinct within a few decades, or perhaps even sooner.

The organic rich A horizon is the living, breathing part of the soil and its location at the very surface is no accident. The fact that the A horizon exists and that it forms where it does is due largely to a single, unusual property of water. Accordingly, the next chapter includes a further discussion of rain, specifically, what happens to raindrops as they fall and why they can only get so large. We also describe how water forms a "wetting front" as it moves into the soil and how it is held near the surface against the force of gravity. Then the narrative focuses on the nature and properties of the A horizon itself, how it forms and why it is so important to life. This will require that we move from the mineral world into the organic world and from the domain of silicon into the domain of carbon.

CHAPTER 6. THE LIVING SURFACE

An earlier chapter described how our planet developed a layered struc-ture as it cooled, with lighter elements coming to the surface and heavier elements remaining behind or sinking to greater depths. Because of this sorting, silicon, aluminum, and oxygen came to dominate both surface rocks and the mineral soils that developed from them. But there is more to soil than minerals; at the very top of most soils is a distinctive, dark-colored surface layer. If you analyzed the organic matter that makes up a significant portion of these layers, you would find very little silicon and very little aluminum. What you would find is a lot of carbon, hydrogen, and oxygen, which is typical of all things that are living or have once lived. The human body, for example, contains just 0.002 percent silicon, while carbon, hydrogen, and oxygen account for more than 90 percent of our body weight, and we are not unique. These same three elements make up most of the plant and animal tissue on Earth. Cellulose, the most abundant organic compound that we know of, is a carbohydrate with the formula $(C_6H_{10}O_5)n$. Lignin, the tough polymer that strengthens tree trunks, has the formula $C_9H_{10}O_2$. This basic C-H-O structure is ubiquitous in animal biology as well. Human cholesterol, for example, is $C_{27}H_{46}O$, while the most common form of fat stored in our bodies is $C_{18}H_{34}O_2$. Testosterone is $C_{19}H_{28}O_2$, while estrogen is $C_{18}H_{24}O_2$. It is no exaggeration to say that just as silicon and oxygen form the backbone of minerals and rocks, the living world is built on a foundation of carbon, hydrogen, and oxygen.

Sister Elements

You probably have read or heard that life on Earth is carbon based, and that statement is true; but what is it about carbon that makes it so important to life? Why not silicon, for example? Carbon is just above silicon in the same column of the periodic table, so the two elements have much in common. Both have four electrons in their outermost shells. Carbon has a total of six electrons, two in the first shell and four in the second (2, 4), while silicon has a total of 14 electrons with a configuration of (2, 8, 4). In order to reach a stable "magic" number, each element must give away those four excess electrons or share them with another atom. A significant difference between silicon and carbon is the way in which they do this. For example, carbon combines with oxygen by way of bonds that are mostly covalent. In contrast, silicon combines with oxygen by way of bonds that are primarily ionic in nature. An ionic bond is one in which an atom largely gives up one or more electrons to another atom, as opposed to covalent bonding, in which electrons are shared in a relatively equal manner between two atoms. It is commonly observed that the living world is carbon based while the mineral world is silicon based, but it also is accurate to say that living things are held together mostly by covalent bonds, while rocks and minerals contain mostly ionic bonds.

Covalent and Ionic Bonds

Linus Pauling assigned electronegativity values to every element in the periodic table, with fluorine ranking as the most electronegative at 4.0 and francium the least electronegative at 0.70. The higher the electronegativity of an element, the more strongly it will pull electrons away from any other element with which it bonds. Consider what happens when sodium, an element of low electronegativity (0.93), comes into contact with a very electronegative element, such as chlorine (3.16). The reaction is spontaneous and violent, but when everything settles down, you are left with very orderly crystals of salt, or $NaCl$. Chlorine largely takes possession of the excess electron that sodium seeks to get rid of, but the electron still remains attached to the Na atom, acting as a kind of bridge holding the Na and Cl atoms together. Salt is a crystal because Na atoms, upon giving away their excess electrons, become positively charged ions (Na^+). Concurrently, when Cl atoms accept electrons, they become negatively charged ions (Cl^-). The + and - forces are of a collective nature, causing the electrical field emanating from each $^+NaCl^-$ to attract a number of nearby $^+NaCl^-$ molecules, while it in turn is affected by others; thus the crystalline structure. A similar process takes place when rocks and minerals form, as oxygen, the second most electronegative element in the world, bonds with elements of lower electronegativity such as silicon. The difference in electronegativity between oxygen and silicon (3.44 vs. 1.90) is not so large, but it still is sufficient to form bonds that are strongly ionic, sufficiently ionic to result in a well-ordered crystalline structure.

Unlike the bonds between silicon and oxygen, those between carbon and oxygen atoms are about 90 percent covalent, meaning that it is mostly a sharing arrangement. One electron from a carbon atom and one electron from an oxygen atom form overlapping orbits which bind the two atoms together much like wrapping a piece of twine around a bundle. A single bond (i.e., sharing one pair of electrons) is rather easy to achieve, but carbon is able to do something that silicon and most other elements cannot do. It can share two electrons with oxygen at the same time, creating what is called a double bond. In essence, each of two different electrons from a carbon atom shares an overlapping orbital with each of two different electrons from an oxygen atom. But there is more. After forming one double bond with an oxygen atom, carbon can then form another double bond with yet another oxygen atom at the same time, resulting in CO_2 or O=C=O. This small, stable, neutral molecule of CO_2 can diffuse freely through the atmosphere with no need to interact or combine with anything else around it; if it encounters another CO_2 molecule or a stray molecule of oxygen or nitrogen, for example, it will simply bounce off and continue on its way. While carbon readily combines with oxygen to form a gas, a combination between silicon and oxygen results in the formation of a crystal, with silicon ions enclosed tightly in an interlocking oxygen matrix extending in all directions. Think of quartz or a massive chunk of granite.

The fact that CO_2 can exist as a gas in the atmosphere and SiO_2 cannot is one reason carbon is so important to life while silicon is not. But carbon has another advantage: Carbon can form stable double bonds (C=C) with itself, something silicon cannot do. The examples below show two kinds of chains that can form when carbon atoms are linked together.

C – C – C – C – C – C – C – C – C – C – C

C – C – C – C – C – C
$$\backslash\backslash$$
C – C – C – C – C –C

In the first chain, single bonds link the atoms together with each carbon sharing a single electron with one or more adjacent carbon atoms. Note that this chain is straight, with no kinks or bends. The second carbon chain has a double bond at the sixth carbon. You will observe that the double bond causes a sharp bend or kink to form in the chain. This is important, because double bonds (and the resulting kinks) at different locations in carbon chains open up a whole world of possibilities when it comes to molecular structure. There is almost no limit to the number of carbon atoms that can link together, and double bonds can be formed at almost any location. This allows for the formation of straight chains, kinked chains, and a variety of branched structures. In addition, strategically placed kinks enable five or six carbon atoms to form closed rings like those that make up the structure of sugars such as sucrose and fructose. Carbon is so versatile that it has been called the "duct tape of atoms" and much of this versatility comes from the kinkiness that results from those simple carbon-to-carbon double bonds ($C=C$).

Why can carbon do the neat double bonding trick with itself while silicon cannot? Much of the answer comes down to size. The covalent radius of carbon (its radius after it has formed a covalent bond with another atom) is 0.67 Å, while that of silicon is 1.11 Å. Because carbon atoms are so small, it is fairly easy for them to form double bonds ($C=C$) by sharing two electrons from each atom. However, silicon atoms are so big, and there are so many other electrons in the way, that it is almost impossible for that second bond to form. Silicon has 14 electrons, while carbon has only six. A double bond occasionally forms between two silicon atoms, but it lasts for only an instant before the two atoms break apart. In contrast, carbon-to-carbon double bonds form easily and are very stable. Largely because of this, they are a crucial part of every living organism on Earth.

Photosynthesis

A simple but revealing experiment is sometimes carried out in classrooms. A snail is placed inside a lidded jar along with a supply of water and a green plant that the snail likes to eat (such as *Elodea*). The snail will begin eating the plant leaves and oxidizing their tissues for energy, generating CO_2 in the process. The plant, in turn, will use the CO_2 released by the snail to grow more leaves, releasing oxygen as a byproduct. The snail will then breathe in the oxygen and eat the new leaves, returning CO_2 to the jar's atmosphere. This cycle will continue for a long time, with the plant depending on the snail for its survival and the snail depending on the plant for its survival. They will reuse the same water, carbon, and oxygen over and over again. If the snail dies, the plant will soon die as well; and if the plant dies, the snail will soon die. Of course, this little experiment will fail if the jar is kept in the dark; sunlight is required to drive the system. If you remove the jar from the sun, both snail and plant will perish. This simple ecosystem, a snail and small plant living cooperatively in a jar, serves as a functional model for the Earth's entire biosphere, in which the same things happen but on a much larger scale. It is a useful demonstration of how the small amount of carbon in the air is continuously recycled in order to sustain life on Earth. The global carbon cycle has been working reasonably well for millions of years, with the CO_2 that plants remove from the atmosphere each year balanced (roughly) by that returned to the atmosphere annually by plant, animal, and microbial respiration.

Gazing at a field of green corn or a lush pasture, you might think nothing much is going on, but you could not be more wrong. In both cases, the seemingly inactive plants are busily absorbing sunlight and taking in carbon dioxide from the air, while beneath the ground millions of tiny root tips are burrowing through the soil in a constant search for water and nutrients. The CO_2 absorbed by the leaves and water taken up by the roots are the raw materials of photosynthesis, while the sun supplies the energy, as shown by the formula: $6\ CO_2 + 6\ H_2O$ + energy from the sun $\Rightarrow$ $C_6H_{12}O_6 + 6O_2$.

In photosynthesis, green leaves absorb sunlight and use the energy to split water molecules into hydrogen and oxygen. The oxygen is given off to the atmosphere as "waste" and the hydrogen is kept. Then carbon dioxide that has come in through small openings on the lower sides of leaves is combined with hydrogen to make $C_6H_{12}O_6$ (glucose). It is clear from a casual inspection that a glucose molecule is larger and has a more complex structure than either CO_2 or H_2O. This order and structure were created by the sun's energy, and plants and animals can recover part of this energy by breaking down or oxidizing the glucose. Here is the equation for the oxidation of glucose: $C_6H_{12}O_6 + 6\ O_2 \Rightarrow 6\ CO_2 + 6\ H_2O$ + energy. This equation is the exact reverse of the photosynthesis equation above. It is important to note that nothing is lost or destroyed in this process. There is exactly the same amount of matter on both sides of this equation: 6 carbon atoms, 12 hydrogen atoms, and 18 oxygen atoms. Count them.

Every green leaf on Earth spends much of its time carrying out photosynthesis and the results on a global basis are somewhat astounding. Each year, green plants remove about 120 gigatons of carbon from the atmosphere. A gigaton is one billion or 10^9 metric tons. Since the total amount of carbon in the atmosphere is only about 750 gigatons, land plants remove about 16 percent of the total world supply each year (120/750). If atmospheric carbon dioxide were not continuously recycled, life on Earth soon would come to an end. This highlights the fact that our global supply of carbon dioxide, the raw material of life, is dismayingly small. To put this in perspective, imagine a town of 10,000 people in which all of the citizens wear sweatshirts reflecting the composition of Earth's air. Since nitrogen makes up 78 percent of the atmosphere, 78 percent of the citizens, or 7,800 of them, would be wearing shirts with a prominent N_2 emblazoned on the front. At the same time, 2,100 citizens would be wearing sweatshirts with a big O_2 logo, reflecting the 21 percent of the atmosphere consisting of oxygen; and perhaps surprisingly, 93 people would be wearing shirts with a big Ar on the front, corresponding to the amount of argon in the atmosphere. Walking around town, you would see a lot of N_2 and O_2 sweatshirts, and even an occasional Ar, but you might wander for days and never encounter a single person wearing a carbon dioxide sweatshirt. Out of the entire population of 10,000 people, only four would be wearing sweatshirts with the CO_2 logo; and just a few centuries ago there would have been only three.

Because the amount of atmospheric carbon is so small and plants remove so much each year, it must be recycled very rapidly. It has been estimated that all of the carbon in our atmosphere is recycled about every 300 years, and much of that recycling takes place in the soil. But let us return to the snail-in-a-jar experiment described earlier. This demonstration is useful in getting a rather difficult concept across to young people, but it does not accurately reflect what happens in the real world (the world outside the jar). Below is a more relevant experiment described by Sir John Russell around 1911 as part of a soil science curriculum he developed for some local schools in England.

> Put some fresh moist garden soil into a bottle and cork it up tightly so that it keeps moist. Write the date on the bottle and then leave it in the light where you can easily see it. After a time—sometimes a long, sometimes a shorter time—the soil becomes covered with a slimy growth, greenish in color, mingled here and there with reddish brown. The longer the soil is left the better. Often after several months something further happens; little ferns begin to grow and they live a very long time indeed. There is at Rothamsted a bottle of soil that was put up just like this as far back as 1874. For a number of years past a beautiful fern has been growing inside the bottle, and even now it is very healthy and vigorous.

At the time Russell wrote this passage, the little ecosystem in a bottle had been recycling the same carbon, as well as the same water and oxygen, for 37 years. If the jar had been left undisturbed, perhaps some little green ferns might be thriving there even today. The Rothamsted experiment is more reflective of real world conditions because it included soil in the process, and soil plays critical roles both in the storage of carbon and in its return to the atmosphere for re-use. Worldwide, soil surface layers store an estimated 1,500 Gt of carbon, about twice the amount present in the atmosphere at any given time. Carbon is continuously added to the soil as plant roots, leaves and stems are recycled and as plants and animals die and their remains are incorporated into the soil. Soil carbon also includes that present in the living bodies of fungi, bacteria, algae, insects, worms, and other invertebrates that live on and in the soil.

There is an area of woodland about 20 acres (eight hectares) in size near my home. The trees growing there are 50 years or more in age and include such species as yellow poplar, white oak, beech, red maple, southern red oak, and loblolly pine. As each summer ends, a great number of leaves and pine needles fall from the trees and form a layer two to more than four inches thick on the ground, often obscuring the walking trails that wind through the area. But by the latter part of the following summer (before the next leaf fall) this thick layer has largely disappeared. There usually is a thin layer remaining at the very surface that has been little changed, but the leaves immediately below are typically skeletal in appearance, with still-intact veins looking like fragile spider webs. Below this layer is a moist, dark brown organic mass in which few remnants of individual leaves are discernible. Deeper still is the top of the soil itself, where leaves and other plant remains from former years have been thoroughly broken down and mixed with mineral particles. In the course of a year, soil organisms have converted the fresh litter laid down the previous fall into a highly decomposed layer less than an inch thick.

The Carboniferous Period

It is intriguing to think about what has taken place in this little forested area over the past century. Each year for the past 100 years, leaves, twigs, and other plant remains have fallen to the ground, forming a layer several inches thick. If this material had not been broken down rather quickly, there would now be an accumulation of plant debris 30 to more than 40 feet (ten to 12 meters) thick covering the entire area. Such a scenario as this—dead plant material falling to the ground and then not decaying—might seem farfetched, but such conditions actually prevailed on Earth for millions of years, during what is now called the Carboniferous Period. This was a time when vast amounts of carbon were taken out of the air by rapidly growing forests, but very little was returned; instead it accumulated on the ground until, in time, much of it was buried by sediments and then stored underground in thick deposits of coal; and this created a problem. Atmospheric CO_2 levels soon were dropping dramatically, while atmospheric oxygen levels began to reach dangerous levels.

The Carboniferous Period lasted about from about 360 million to about 300 million years ago. During that time, much of the landmass that would become North America was located near the Equator, and shallow inland seas covered a large part of what we now call the United States. The climate was mild and uniform, constant summer to spring-like conditions under which luxurious jungle-like vegetation covered the landscape. Giant horsetails grew profusely along the edges of swamps, while the tree-like ancestors of modern ferns grew to heights of 50 feet or more. Lepido-dendron, an early ancestor of much smaller modern plants called ground pine, dominated the forest canopy, soaring to heights well over 100 feet and measuring as much as five feet in diameter.

These primordial trees were able to grow tall because of lignin, a tough, carbon-rich polymer that is believed to have evolved some-time around 400 million years ago. Lignin gives tree trunks their great strength, enabling them to loft foliage high above the ground to compete for sunlight. We know how large some of these ancestral trees were because parts of many of them have been preserved in coal seams, and how and why this took place makes for an interesting story. When early forest giants died and fell to the ground, no fungal or microbial species then living was able to fully break them down and consume the wood. Lignin was the problem. Thousands of generations of these forests flour-ished and died; and upon falling to the ground, lay mostly inert, resistant to decay. In time, much of this woody debris was buried by sediments and slowly converted into coal. As a result, carbon was taken out of the atmo-sphere in huge amounts and stored underground and thus was removed from the carbon cycle. The amounts are somewhat staggering. We now know that about 20 feet of original plant material must be compressed to produce a coal seam one foot thick. I have seen single exposed coal seams in southern West Virginia and eastern Kentucky more than four feet thick, and some Pennsylvania coal beds (multiple seams) are more than 200 feet (60 meters) thick (Farb 1963).

The good news is that the large deposits of coal being laid down were rich in energy, energy that later would fuel the Industrial Revolution. The bad news was that very little carbon was being returned to the atmo-sphere, so carbon dioxide levels were crashing; at the same time, oxygen levels in the atmosphere had soared to an estimated 35 percent, compared to the current level of around 21 percent. Low levels of carbon dioxide and high levels of oxygen made it increasingly difficult for plants to grow and, in addition, were setting the stage for a rapid descent into an ice age. But then something happened: About 300 million years ago, the once-resistant woody debris began to decompose and the age of coal formation came to a rather abrupt end.

More than 70 coauthors recently published a paper with the somewhat intimidating title, *The Paleozoic origin of enzymatic lignin decomposition reconstructed from 31 fungal genomes* (Floudas et al. 2012). Based on genetic studies of fungal evolution, the authors showed convincingly that white rot fungi, the only organisms on Earth able to break down lignin effectively, came into existence at roughly the same time that the coal age began to end; and there is reason to believe that this was not a coincidence, that these fungi played a major role in bringing the Carboniferous Period to a halt. Here is how it probably happened. Upon evolving the ability to break down lignin, white rot fungi became so successful that their spores soon were spreading throughout Carboniferous forests, finding rich sources of food and energy in the dead wood that had accumulated in such quantities. The fungi made quick work of it, bombarding the now-vulnerable wood with highly reactive oxygen molecules, reducing it to cellulose-rich slurry that could be easily digested. Some have described this crude but effective process as analogous to untangling a knot in a rope by using a blowtorch.

The evolution of white rot fungi had at least two major effects on the world. First, it started the modern carbon cycle by breaking the impasse that had lasted for millions of years, allowing carbon to flow back to the atmosphere. But white rot fungi carried out only the first stages of organic matter breakdown. Bacteria and other organisms had to finish the job, and there had to be a place for this to happen. This place was the soil A horizon, which began to evolve shortly before 400 million years ago. Therefore, in addition to starting the modern carbon cycle, the rise of white rot fungi also set the stage for the evolution of modern soil profiles. Thin, vestigial A horizons probably had existed much earlier. Long before forests evolved, the remains of bacterial crusts, which appeared as early as 2.6 billion years ago (Watanabe et al. 2000), were incorporated into the mineral surface; but so little carbon was in play that the overall effects on soil properties or on the carbon cycle would have been rather small.

The rise of forests and the evolution of white rot fungi changed things dramatically, releasing a lot of carbon into the environment and setting the stage for organic-rich soil surfaces to form and to play a major role in the cycling of carbon through the biosphere. Here is a summary of how this works today. During the growing season, green plants, especially trees, take up large amounts of carbon from the atmosphere as CO_2 and convert it into new tissue. Plants absorb about 120 gigatons of carbon from the atmosphere annually. About half of this (60 gigatons) is returned to the atmosphere fairly quickly by way of plant respiration, while the other 60 gigatons are used for new growth—to make leaves, stems, roots, seeds, and other organs.

The fact that vegetation takes so much carbon out of the air each year and immobilizes so much of it would be a problem were it not for soil. Fortunately, while plants are busily growing above ground, organisms such as fungi and bacteria are hard at work in the soil, breaking down leaves, stems, roots, and even entire dead plants that fell to the ground in past years. The ultimate products are CO_2 and H_2O. Worldwide, this process returns about 60 gigatons of carbon to the atmosphere annually, essentially replacing that stored in new plant tissue each year. Because of the continuous recycling of CO_2, the amount of carbon in soil, air, and vegetation remains fairly constant. As shown in the table below, the soil stores much more carbon than either the atmosphere or land vegetation.

Table 6.1. Amounts of carbon stored in the atmosphere, in land vegetation, and in soil

Data from Smil (1997)

Location	Gigatons (10^9 tons) of carbon
Atmosphere	750
Land plants	600
Soil	1,500

The present-day value of 750 gigatons of atmospheric carbon is somewhat higher than levels that occurred just before the onset of agriculture and modern industry. Worldwide, soils contain an estimated 1,500 gigatons of carbon at any given time, about twice that stored in the atmosphere or in living plants. About 90 percent of this is in the form of dead organic matter, while about ten percent is in the form of live plant roots or in the bodies of living bacteria, fungi, and other soil organisms. Average residence time for carbon in soil is about 30 years, ranging from about ten years or fewer in tropical grasses to 500 years or more for tundra and peat bogs. Because some plants can live to be hundreds or even thousands of years old (in the case of the longest-living trees), some carbon is taken out of circulation for relatively long periods of time.

The ultimate fate of almost every plant on land is to be consumed by an herbivore or to die, fall to the ground, and then be decomposed by the smaller but more numerous organisms living there. Softer, moister tissues, such as leaves and flowers, are broken down quickly, while more resistant structures such as dense wood persist much longer in the environment. But in the end, few living things escape the efficient engines of decomposition that modern soil surfaces have become. Once white rot fungi had made it possible for lignin to be broken down efficiently, an abundant source of food and energy became available at the soil surface. As a result, this zone soon was teeming with life, populated by untold species of microbes, fungi, insects, worms, and even small reptiles and mammals. This dynamic, living surface began to play increasingly vital roles in the economy of nature. At the same time, soils for the first time took on a truly modern look.

The A Horizon

Sir John Russell (1957) described what a typical British gardener might see if he or she dug down a few feet and examined the soil.

> He [or she] will find a fairly well marked distinction between the top layer, which may be 6 to 12 inches in depth, and the lower layer. The top is darker in colour, less sticky, looser, crumb-like in structure... The reason for the difference from the subsoil is that the organic matter which is necessary to complete the soil... has come from the soft roots of grasses and herbaceous plants and from plant residues falling on the surface and carried down into the soil by worms and other animals...and none of these causes usually operates at more than about 6 to 12 inches below the surface.

The last phrase in the quotation above bears repeating: "None of these causes usually operates at more than about 6 to 12 inches below the surface." Although the statement is a little oblique, it makes an important point: the processes that create living, organic-rich A horizons in soils are active only to a shallow depth, rarely extending more than ten inches (25 centimeters) or so below the surface. There are notable exceptions; for example, some soils of the American Midwest once occupied by tall grass prairie have dark surface horizons that are two feet (60 centimeters) or more in thickness. The same is true of some forest soils that formed under very wet conditions. But in general, if you were to examine soils from around the world, you would find that the A horizon, commonly referred to as topsoil, typically ranges from five to about ten inches in thickness.

It might seem that soil surfaces are too thin and inconsequential to be of any real significance, but think of the ozone layer, which also is essential to terrestrial life. Out of every million molecules in the atmosphere, only ten or so are ozone; if all of the ozone in the atmosphere were compressed into a single layer at the earth's surface, it would be only three millimeters or about one eighth of an inch thick. When compared with ozone, the soil surface layer is quite substantial. Although different in many respects, the ozone layer and the soil surface have at least two things in common: Both are essential for life to thrive on land, and both have the capacity to renew or replenish themselves. As ozone molecules are destroyed, more are created; and as the soil surface is slowly eroded away, it is replenished by weathered rock from below and by the addition of organic matter from above.

Soils covered in vegetation erode very slowly, perhaps only a few centimeters or so every 50 to a 100 years. At the same time, rock material from below is slowly but inexorably being broken down into ever-smaller pieces. As the soil slowly eats its way into the landscape, more rock fragments come within the reach of plant roots and are broken down into even finer particles, gradually turning from rock into mineral soil. While this is happening underground, another kind of soil formation is taking place at the surface. Blades of grass, tree leaves, twigs, and even entire trees that fall to the ground are attacked and devoured by an army of fungi, bacteria, worms, insects, and other organisms. As this process continues, decomposed organic matter begins to coat mineral particles and become thoroughly mixed with them, creating the distinctive dark surface layers so typical of vegetated soils everywhere. But why does most of the soil organic matter remain in a thin layer at the surface instead of being dispersed to greater depths? In order to answer this question, it is necessary to consider the nature of water, how it falls from the sky as rain, and how it moves into the soil.

Rain and How It Falls

Chapter 2 introduced the four fundamental forces of nature—the weak nuclear force, the strong nuclear force, gravity, and electromagnetism. The first two operate within the nuclei of atoms, so their effects are not readily apparent to us in our everyday lives. But we have no trouble seeing and feeling the effects of gravity and electromagnetism in the world around us; they affect nearly everything that goes on. How raindrops fall through the air and how water moves into the soil are especially revealing examples of the unending struggle or tug of war that goes on between these two great forces.

Let us begin by considering the nature of rain, and a good place to start is by introducing an intriguing character named Wilson A. Bentley, who lived more than a century ago. Bentley is best known for the thousands of beautiful pictures of snowflakes he took over many years. Atmospheric scientist Duncan Blanchard published a biography of Bentley in 1998 titled *The Snowflake Man.* Wilson Bentley was a short (only about five feet tall), reclusive man who lived all his life in the Vermont farmhouse in which he was born. Although small in stature, he was reportedly very strong as a result of a lifetime of farm work, such as milking cows, cutting hay, and digging potatoes by hand. Largely self educated, he did not attend school until age 14 and then only intermittently. Despite a lack of formal education, his 1904 paper, *Studies of Raindrops and Raindrop Phenomena,* was cited as an extraordinary article introducing new ideas and concepts on raindrop size and formation (Blanchard 1998).

Bentley had long been obsessed with snowflakes, but in 1898 he developed an avid interest in rain, an interest prompted by, in his own words, "a desire to add, if possible, a little to our knowledge regarding rainfall phenomena." During the next six years or so, he obtained 344 measurements of raindrop size from 70 different storms. In order to measure the size of raindrops striking the ground, he devised a simple but effective method (Bentley 1904).

> The method employed was to allow the raindrops to fall into a layer...of fine, uncompacted flour...The raindrops were allowed to remain in the flour until the dough pellet that each drop always produces at the bottom of the cavity was dry and hard. These dough pellets were found by careful experiment to correspond very closely in size with the raindrops that made them.

Bentley's data showed that even in the most torrential storms, raindrops rarely exceeded about six millimeters (one-fourth inch or so) in diameter. Mathematical calculations later showed that very heavy rain will produce some drops as large as five to six millimeters, but they are so rare that only one drop that large is found in ten cubic meters of air (Blanchard 1966). Even in the most torrential rains, the most common raindrop size is three millimeter (about one-eighth inch) or less. Bentley's work was largely ignored for nearly 40 years, but in 1943, two scientists working for the Soil Conservation Service, J.O. Laws and D.A. Parsons, read his 1904 paper and used his technique to measure raindrop diameter. They were interested in raindrop size and its effect on the intensity and energy of rainstorms as a way of predicting rates of soil erosion. They found that a heavy rain produces much larger raindrops than a light rain and that these larger drops fall much more rapidly. The fact that larger raindrops fall with greater velocity than smaller drops is of great importance, because the kinetic energy of a raindrop hitting the ground is proportional to its mass (m) and its velocity squared (v^2).

But why does a diameter of five to six millimeters (one-fifth to one-fourth inch) appear to be the upper limit for raindrop size? Why, even in very intense storms, is the median raindrop size about three millimeters (one-eighth inch)? To answer those questions, let us turn to German scientist Philipp Lenard, who became interested in rain at about the same time as Wilson Bentley. Two men could hardly be more different. Bentley was a life-long farmer with little formal education, while Lenard was an experimental physicist with advanced degrees who taught at some of the leading universities in Europe. In 1904, the same year that Bentley published his work on rain, Lenard published an article titled *Uber Regen* (*About Rain*) in *Meteorologische Zeitschrift* (*Meteorological Magazine*). He had constructed a vertical wind tunnel in which the upward speed of the air could be adjusted to match the velocity of the falling drops, causing them to float in the air. One important thing his studies did was to dispel the myth of the tear-shaped raindrop. Lenard found that falling drops of water start out as small glistening spheres, but eventually they flatten out from the pressure of the air below, assuming the shape of a bagel or fat hamburger. Like Bentley, Lenard found that raindrops almost never exceeded about six millimeters in diameter. He was awarded the Nobel Prize in 1905 for earlier work he had done on cathode rays, and like Bentley, never published another paper on rain (Blanchard 1972).

As suggested by Lenard's wind tunnel experiments, a falling raindrop is a very good example of the struggle between gravity and electromagnetism. The polarity of water molecules causes them to bond together, with the slightly negative (oxygen) end of one molecule attaching itself to the slightly positive (hydrogen) end of another. As a result, water is "sticky" at the molecular level, giving it an unusually high surface tension, so high that small insects can walk on the surface of a still pond. This same surface tension causes water to be held by small pores in the soil, thus creating a "hanging water table" at the surface, storing water after each rain to be taken up later by thirsty plant roots.

This hanging water table that forms at the soil surface is of great importance to terrestrial life. In order to understand this phenomenon, let us return to the subject of raindrops and what happens as they fall from the sky. As discussed earlier, in order for a raindrop to form, many tiny water droplets must condense around a microscopic particle of salt, dust, or smoke. This drop then must grow large and heavy enough so that its fall speed is greater than the updrafts of air keeping the clouds aloft. As small raindrops fall down through the clouds, they pick up additional water, much like snowballs rolling downhill. The polarity of water molecules causes them to stick together, initially shaping the small drop into a sphere. But as the raindrop grows larger and falls faster, it encounters more air resistance from below, which causes the bottom to flatten out. As they flatten out, very large drops are more likely to break apart than small drops. Bentley and Lenard both concluded that the upper limit to raindrop size is about five to six millimeters (about one-fifth to one-fourth inch) and that the typical raindrop size is about half this size. Raindrops much larger than five to six millimeters in diameter flatten out to the point that they become unstable and break apart.

We are fortunate that gravity, atmospheric density, and the attractive forces between water molecules all fall within a Goldilocks zone, ensuring that a sufficient number of raindrops grow large enough to start falling from the sky, but not too large. Gravity and electromagnetism interact in such a way that raindrops are not unduly large and normally fall to the ground with only moderate force and intensity, so that most rain striking the soil surface is easily absorbed. During the growing season, nearly all of the rain falling on forests, on grasslands, or on well-managed cropland, if not intercepted by leaves, moves down into the soil, with very little flowing across the surface to cause erosion. Very intense rain events with very large raindrops are rare. Allen (1983) emphasized how important the small particles of salt, dust, and ash that form the nuclei of raindrops are. "If no such particles existed in the air, clouds and rain would have a completely different character. The air would become supersaturated with vapor, without any clouds appearing; then from time to time massive clouds would form abruptly and ruinous downpours would lash the earth. Gentle rains would almost never occur."

Wetting Front

When it comes to rain and the way it forms and falls from the sky, nature has been very kind to us. We also are fortunate in the way rain moves into and through the soil. Black (1957) describes what happens in the upper part of the soil after a significant rain event. "During entry of water into a dry soil the water moves downward in a "front," across which the water content changes in a short distance from that of the moistened soil to that of the initial water content of the dry soil. When water intake ceases, penetration of the front continues, but usually its movement has almost ceased within 2 to 3 days."

Black's description reflects the experience of both Russian and American scientists who, early in the 20th century, closely observed what happened when a measured amount of water (such as a one-inch rainfall) fell onto a soil surface. Veihmeyer (1927) measured the water content at small depth intervals in a soil two days after a little over two inches of rain had fallen. He found that the soil was moistened uniformly to a depth of about 14 inches (35 centimeters), and then there was an abrupt transition to the much drier soil below. The upper moistened layer contained 22.9 percent water by weight, while the unwetted soil below contained only 9.4 percent moisture. Shaw (1927) performed a similar experiment. Adding six inches of water to the surface of a column of dry soil, he observed that the soil was moistened uniformly to a depth of 24 inches (61 centimeters), at which point there was an abrupt transition to dry, unaltered soil below. In 1897, Briggs had proposed that water near the surface was held in small pores by capillary forces strong enough to resist the downward pull of gravity. He surmised quite correctly that soil pores act much like capillary tubes. As pores become smaller, hydrogen bonding between water molecules and between water molecules and the walls of pores slow down movement. Eventually a hanging water table is created as water occupies all of the pores small enough to hold it against the downward pull of gravity.

The porous, organic-rich soil surface acts much like a sponge, soaking up and retaining water, creating a wetting zone that is recharged (totally or in part) each time rain falls to the ground and percolates into the soil. This zone in humid regions typically extends from the surface to a depth of five to ten inches (12 to 25 centimeters). Fungi, bacteria, and other soil invertebrates thrive in this zone, which, in addition to providing life-enhancing water and an abundant supply of organic carbon, also provides a living space in which the above ground climatic extremes are dampened and moderated. Not surprisingly, the feeding (and drinking) roots of higher plants also are concentrated at the soil surface because this shallow layer is where water is most frequently replenished. This enables the growing plant to get the maximum benefit from each rainfall event. The fact that a relatively thin layer at the soil surface is replenished with water after every rain is critical to the survival and growth of land plants, because during photosynthesis, "Several hundreds of pounds of water may be lost by translocation during the production of a single pound of dry matter" (Black 1957).

Structure and Porosity

Soils containing more organic matter are looser, friable, and more porous, with many open spaces through which both water and air can readily move. This, along with the insulating properties of organic matter, creates a rich, nurturing environment for the multitude of creatures that make their home in the soil. Soil surfaces made friable and porous by the presence of organic matter are typically described as having good structure, meaning that individual particles are loosely clumped together into discrete bodies, giving the soil a crumbly appearance when gently broken apart. A surface soil with good structure might look like the inside of a very rich chocolate cake or like a mass of coarse coffee grounds. You can see some good examples of this by keying a phrase such as "pictures of soil structure" into a search engine.

Soil develops structure as a result of small mineral particles being clumped together to form larger units. The biological "glues" that do this include such things as decomposed and decaying organic matter, fungal mycelia, root hairs, and the mucilaginous remains of worms and other soil animals, with clay and iron oxides often playing prominent roles. When some of these binding agents dry, they tend to shrink, pulling the bound particles together into aggregates. Distinct structural aggregates are typically about five millimeters (one-fifth inch) or smaller in diameter. Such aggregates can persist for some time, and even if destroyed, are soon re-formed in healthy soils.

In general, the longer a soil surface remains undisturbed and covered in vegetation, the better the structure. A British study found that very good structure developed in a sandy soil under grass in as little as ten years (Low 1955). In contrast, 50 years or more were required for similar structure to develop in clay soils. Good structure optimizes root growth and the movement of water and air into and through the soil, resulting in greatly enhanced plant growth. Rynasiewicz (1945) found that the correlation between the percentage of soil aggregates and the yield of onions was 0.99, almost a perfect straight-line relationship. The crumbs or aggregates typical of good structure are irregular in shape and do not fit closely together. In addition, the aggregates themselves are not solid throughout; instead they are porous and able to absorb water like tiny sponges. A clod of soil might appear solid to us, but half or more of its volume consists of empty space—well, not really empty, since any void in the soil is always filled with air or water.

Organic Matter and Water

In 1931, Veihmeyer and Hendrickson defined what they called field capacity as "the amount of water held in the soil after the excess gravitational water has drained away and after the rate of downward movement of water has materially decreased." As already indicated, the water enters small pores where it is held by capillary forces or is held in thin films around soil particles. As roots begin extracting water, the pores begin to empty and water films around particles become thinner and thinner, until eventually the soil holds the remaining water with such force that roots cannot extract it. At this point, plants begin to wilt. Not surprisingly, the water content at which this starts to take place is called the wilting point. The amount of water held at field capacity minus the amount held at the wilting point is referred to as plant-available water. Table 6.2 shows the amount of water held at field capacity and the plant-available water for a sand surface layer and for a silt loam surface layer with different amounts of organic matter. The units are inches of water in the top 6 inches (15 centimeters) of soil.

Table 6.2. Water held at field capacity and plant-available water (inches of water held in 6-inch thick surface layer) as affected by texture and organic matter content

Adapted from Hudson (1994)

	Water held at field capacity	Plant available water
Sand, 1% organic matter	0.50	0.40
Sand, 4% organic matter	1.00	0.75
Silt loam, 1% organic matter	1.40	0.80
Silt loam, 4% organic matter	2.20	1.45

A layer of water one inch thick spread over an entire acre (sometimes referred to as an acre-inch) is equivalent to 27,154 gallons, so it is easy to convert inches of water in the soil to gallons per acre. The sand surface layer with one percent organic matter, when at field capacity, holds about 14,000 gallons of water per acre, while the silt loam surface with the same amount of organic matter holds 38,000 gallons, nearly three times as much. The amount of water available to plants also is much higher in the silt loam, about 22,000 gallons per acre versus about 11,000 gallons per acre in the sand. A silt loam soil can supply much more water to crops or to natural vegetation than a sandy soil, so in nearly all cases plants will grow much better on silt loam than on sand; so texture really matters; but so does the amount of organic matter present. In both textures, the amount of water held at field capacity and the amount available to plants nearly double when organic matter increases from one percent to four percent by weight. The agricultural and ecological implications of this are obvious: Whatever kind of land is being farmed and whatever the soil texture, one should aim to maintain and, if possible, increase the organic matter content.

The fact that more organic matter increases the porosity of a soil and enables it to store more water from each rain and make more of it available to plants is important because photosynthesis is extremely wasteful when it comes to water. During photosynthesis, CO_2 is absorbed through millions of tiny openings on the underside of leaves called stomata (from the Greek *stoma*, or mouth). Stomata must remain open in order to capture CO_2, and in the process a lot of water escapes. Even the most efficient plants can lose 500 or more molecules of water for each molecule of CO_2 taken from the atmosphere. As water is continuously lost, it must be replaced from the soil below; otherwise the plant will wilt and eventually die. That is why the capacity of the soil surface to take in water, store it, and then supply it to plant roots is so important — and that is where organic matter plays such an important role.

You might wonder why hypothetical soil surfaces 6 inches (15 centimeters) thick were chosen for the analyses above. This particular thickness was chosen for a number of practical reasons. Six to seven inches is the approximate depth to which farming implements such as plows, disks, and harrows commonly disturb the soil. It also is the layer in which most of the water-absorbing roots are concentrated. This is true not just for crop plants but for native forests and grasslands as well. The anchoring roots of a large tree can be quite deep, but most water is taken up by the profusion of small roots concentrated near the soil surface. Water-absorbing roots are concentrated in the top 6 to 7 inches (15 to 18 centimeters) or so because that is the wetting zone, the layer of soil that typically is recharged with water every time it rains. There are a few notable exceptions, such as certain desert plants, but most forest and grassland species adopt a "concentrate the water-absorbing roots near the surface" strategy. Note: Europeans use a depth of 15 centimeters (6 inches) for their hectare-furrow-slice. Historically, American agronomists have used a depth of 6.7 inches for their acre-furrow-slice. For ease of calculation, this book follows the European convention.

Some might question how such relatively small amounts of organic matter, one to four percent by weight, can have such profound effects on water retention. The key phrase here is "by weight." Organic matter has an average bulk density of about 0.3 to 0.5 gram per cubic centimeter (g/cm³), so it is very light in comparison to mineral soil, which most often has a bulk density of 1.50 g/cm³ or higher. Because of this large difference in density, one percent organic matter by weight is about five percent on a volume basis, while four percent organic matter by weight is about 20 percent by volume. Water stored in soil accounts for a significant portion of the terrestrial supply of fresh water. At any given time, the world's soils contain about 1.3 times as much water as the atmosphere, nearly eight times as much as all of the world's rivers, and nearly 15 times as much as in the tissues of all living terrestrial plants and animals (Shiklomanov 1993), and much of this storage capacity (typically half or more) is due to the presence of organic matter.

Gravity, Electromagnetism, and Pore Size

Some pores in the soil are so large that water moves freely through them, evidence that gravity is the dominant force. But other pores are so small that any water entering them is held firmly. If soil pores are larger than about 75 microns (µ), water drains out of them readily. For purposes of scale, the average human hair is about 100 µ in diameter. Here are the critical pore sizes that determine whether water is retained by the soil or quickly drains away under the force of gravity (adapted from Brewer 1964).

- Pores larger than 75 µ (about 3/4th the thickness of a human hair) will not retain water against the force of gravity; instead, it drains away rapidly.

- Pores between 30 µ and 75 µ in diameter (about 1/3rd to 3/4th the thickness of a human hair) will hold water for a short time, but if not taken up by plant roots, the water gradually drains away.

- Pores smaller than 30 µ (about 1/3rd the thickness of a human hair) will retain water almost indefinitely against the force of gravity.

However, this water is easily accessed by root hairs, which are rarely larger than about 15 µ in diameter.

The addition of well-decomposed organic matter to the soil surface creates a lot of pores in the 30 µ to 75 µ range, increasing the amount of water available to plant roots immediately after a rain. This allows roots to harvest the water before it drains away. Between rain events, roots hairs can extract the water from long-term storage in pores that are about one third the thickness of a human hair or smaller. In addition to increasing the water-supplying capability of soil, organic matter has other positive benefits.

Organic Matter and Electrical Charge

As described earlier, cation exchange capacity, or CEC, refers to the total number of electrical charges in a given amount of soil. Historically, it has been expressed in milliequivalents per 100 grams (me/100g). It is important to remember that one milliequivalent is equal to 6.022×10^{20} electrical charges, a very large number. Just 100 grams of kaolinite clay with a CEC of 10 me/100 grams would have approximately 6,022,000,000,000,000,000,000 negative electrical charges on its surfaces. But montmorillonite has ten times as many. This brings us to soil organic matter, which is even more electrified, with twice the electrical charges of montmorillonite and 20 times that of kaolinite. As discussed earlier in the book, the more electrically charged a soil is, the more ions such as Ca^{2+}, Mg^{2+}, and K^+ it can hold on its surfaces and make available to plants, and this is where organic matter excels. A surprisingly small amount of organic matter can greatly increase the total number of electrical charges in the soil.

The electrical charges on soil organic matter act much the same way as those on clays, but they have a totally different origin. According to Bohn et al. (1979), "The negative charges of the soil organic fraction is generally agreed to be due to the dissociation of H^+ from certain functional groups, particularly carboxylic and phenolic groups." This statement refers to a rather straightforward process. Functional groups are common in biological systems and are characteristic of certain kinds of molecules. For example, the hydroxyl functional group (¯OH) is characteristic of alcohols, while the amino functional group (¯NH$_2$) is characteristic of amino acids. Carboxylic groups (¯COOH) are prominent in the structure of both proteins and fats, so it should not surprise us that large numbers of them end up as part of soil organic matter, which is derived from once-living plants and animals.

The carboxylic group, along with the phenolic group (basically a benzene ring with an OH attached), accounts for 85 to 90 percent of the electrical charges on soil organic matter. To illustrate how these charges come about, let us start with an exposed carboxyl group. The COOH or COO⁻ H⁺ acts like a weak acid, readily giving up its H⁺ in the presence of OH⁻ ions. The negative charge on the exposed COO⁻ can then attract a positively charged ion, such as Ca^{2+}, Mg^{2+}, K^+, or Na^+.

$$COO^- H^+ + OH^- \Rightarrow COO^- + H_2O$$
$$COO^- + K^+ \Rightarrow COO^- K^+$$

At any given time, a single gram of soil organic matter can have trillions of exposed functional groups and the overall result can be quite dramatic. Remember that a single milliequivalent of CEC is equal to more than 600 trillion electrical charges. Helling et al. (1964) determined how many of the electrical charges in 60 Wisconsin soils were due to clay and how many due to organic matter. They found that, on average, organic matter accounted for 70 percent of the electrical charges, while clay accounted for 30 percent. The overwhelming effect of organic matter on CEC is not unique to these soils or to this region. Organic matter has similar or even greater effects in other parts of the United States and in other parts of the world. In some Ultisols of the southeastern United States and in many Ultisols and Oxisols of Africa and South America, organic matter probably accounts for significantly more than 70 percent of the electrical charges.

This chapter has described three important functions of soil organic matter. First, it increases the amount of water a soil can store and the amount it can make available to plants. At any given time, half or more of the water in soil surfaces worldwide is due to the presence of organic matter. Second, organic matter increases the number of electrical charges in the soil. Half or more of the electrical charges in soils worldwide are the result of exposed functional groups on the surfaces of organic matter. As a result, half or more of the plant-available calcium, magnesium, and potassium present in the world's soil surfaces is held there because of organic matter. Third, soil organic matter plays a crucial role in the carbon cycle, which replaces all of the CO_2 in the atmosphere about every 300 years. By storing large amounts of carbon and releasing it slowly, soils help to keep the atmospheric levels of CO_2 stable over time.

But just to be clear, the effect of organic matter on the carbon cycle (or on any other cycle) is an indirect one. Soil organic matter merely serves as an energy source and as a carbon source for the vast array of organisms that actually do the work. The total numbers of these organisms are almost beyond comprehension. Below is a partial listing of the numbers and kinds of organisms one might find on just one acre of rich grassland.

- Trillions of bacteria
- Billions of fungal bodies
- Millions of miles of fungal mycelia
- Hundreds of millions of protozoa
- Tens of millions of nematodes
- Millions of algae
- Thousands of earthworms

If this many soil organisms can be found in a single acre, just think of the staggering numbers there must be worldwide—which brings us to yet another reason why soil organisms, especially soil microbes, are so important; they act as a biological storehouse for the world's essential genes.

The Engines of Life

Biologist Paul Falkowski (2015) described what he calls biological nanomachines, or "assemblies composed largely of proteins and nucleic acids and carry out the necessary functions of all living things." These nanomachines operate entirely within cells, so in addition to being present in the cells of so-called higher animals and plants, they also are found in single-celled organisms such as bacteria. Falkowski argued that there is a surprisingly small number of these essential nanomachines, that they originated very early in Earth's history, and that they have been highly conserved over time. When a biological function, molecule, or structure has been around for a long time and is the same or very similar across many life forms, biologists say that it is highly conserved. This means that, even as organisms adapt to new conditions through natural selection, certain things that are essential to survival remain relatively unchanged despite many generations and many evolutionary challenges.

Perhaps the simplest way to explain this is to describe two of nature's nanomachines that are both highly conserved and essential. By "essential" I mean that without them, 99.9 percent of the biomass now on Earth would not exist. One of them, a protein called photosystem II, which is nature's water-splitting enzyme, is so important that one biologist (Barber 2003) has referred to it simply as the "engine of life." Below is the now familiar equation for photosynthesis, in which the sun's energy is used to convert carbon dioxide and water into glucose and oxygen.

$$6\,CO_2 + 6\,H_2O + \text{energy from the sun} \Rightarrow C_6H_{12}O_6 + 6O_2$$

But before any of this can happen, water molecules must be broken down into their component parts, as shown here.

$$2\,H_2O \Rightarrow 4H^+ + O_2 + 4e^-$$

The oxygen is released to the atmosphere, enabling animals such as human beings to exist, while the H^+ ions are combined with CO_2 to make glucose, which nearly every life form on Earth uses as a source of energy. But without photosystem II to act as a catalyst, water molecules could not be split apart and photosynthesis could not proceed. Photosystem II is believed to have evolved just once, around 2.6 to 2.7 billion years ago, and has been passed down relatively unchanged for many thousands of generations. If you took a sample of this protein from a leaf of a plant growing today and compared it with a sample taken from a cyanobacterium that lived more than a billion years ago, you would find that they were remarkably similar.

Another nanomachine, called the coupling factor, is essential to respiration, which, as discussed earlier, is the reverse or opposite of photosynthesis. In photosynthesis, H^+ ions are stripped from water molecules and combined with CO_2 to make glucose. In respiration, the H^+ ions are stripped from the glucose, used to generate energy, and then reunited with oxygen to form water again. In essence, photosynthesis splits water molecules, while respiration puts them back together again. Below is the equation for respiration.

$$C_6H_{12}O_6 + 6O_2 \Rightarrow 6\,CO_2 + 6\,H_2O + \text{energy}$$

This is a very precise equation except for one thing; the word energy at the end is very vague. What kind of energy are we talking about and how much? Fortunately, we can be more specific. For each molecule of glucose ($C_6H_{12}O_6$) that is oxidized, 38 molecules of ATP are created. ATP is the energy currency of all life, and the coupling factor is a little ATP-producing machine. This nanomachine evolved in microbes long ago and now has spread to all of the kingdoms of life; and it has essentially the same structure and works the same way in all of them. The coupling factor is not just a machine in the metaphoric sense; it actually looks like a machine. It has the appearance of a tiny merry-go-round, with its central shaft passing through a cell membrane; it makes ATP by ferrying protons (H^+ ions) across the membrane. The H^+ ions, driven by the energy of electrons, move along the shaft. In the process they pass through the membrane while causing the entire apparatus to rotate counterclockwise. On average, three

ATP molecules are formed and released into the cell with each complete revolution of the "merry-go-round."

Every new generation of living things must create anew the two essential nanomachines just discussed and others as well. This means that the genetic codes, which are the instructions for doing so, must be passed on. Instead of thinking in terms of essential nanomachines, it probably is more accurate to think in terms of the essential genes that resurrect them generation after generation. With that in mind, just how many essential genes are there? Falkowski estimates that there are about 1,500. This is a very small percentage of the 60 to 100 million genes believed to exist in the world. He cautions that his estimate could be too conservative, and the true number might be closer to 15,000, but this still is a relatively small number, about 0.015 percent of the total.

The small numbers of genes that are essential to life have managed to survive for more than two billion years despite numerous worldwide extinction events. The most recent one occurred about 65 million years ago when a large meteor struck near Mexico's Yucatan peninsula. This catastrophic event is believed to have doomed the dinosaurs and many other plant and animal species; but according to Falkowski, "Microbes sailed right through that extinction, just as they had in all the other previous extinctions that span vast times in Earth's history." Four such events are known to have occurred during the last 550 million years, with each leading to a catastrophic loss of species, but the world's microbes were little affected by any of them; and that is a clue to how nature protects its essential genetic material.

Every essential gene is housed in a wide variety of microbes scattered throughout the world. Such important work could not be entrusted to multi-cellular organisms such as humans; we are far too few and far too fragile. Microbes are much better suited to the task. They are tiny, can be sustained in vast numbers, and can rapidly colonize every nook and cranny on Earth that receives even occasional moisture. There are many trillions of microbes, with especially large numbers in or on the soil, where they constitute a vast genetic reserve. What kind of natural disaster would it take to destroy all of the microbes in all of the world's soils to a depth of, in some cases, many meters? Short of blowing the planet up, it is hard to imagine one. Unfortunately, if soil microbes (and the essential genes they harbor) are ever called upon to resurrect life on our planet, no human will be around to see it, since we already will have perished. We will never know if nature's microbial insurance policy has paid off.

CHAPTER 7. PARENT MATERIAL AND CLIMATE

Soils cover most of the world's land surface and as environmental factors vary across this surface, so do soils. Hans Jenny, in his classic 1941 book, *Factors of Soil Formation*, examined the nature of this variation, showing conclusively that the five factors listed below are the "independent variables that define the soil system":

Climate (cl)
Organisms (o)
Topography or relief (r)
Parent material (p)
Time (t)

He wrote, "For a given combination of cl, o, r, p, and t, the state of the soil system is fixed; only one type of soil exists under these conditions." In essence, the soil at any location on Earth is a function of climate, organisms, topography, parent material and time, or in symbolic form: s = ƒ (cl, o, r, p, t). The letters representing the factors of soil formation are sometimes written without the commas, yielding *clorpt*, creating a useful mnemonic or memory aid. Jenny argued quite convincingly that identifying the soil forming factors enables one to predict the kind of soil to be found at any location in the world. We will explore this idea by considering some examples from different areas of the United States. But doing this requires that we have identifiable kinds of soil to use as the objects of our study, which brings us to Soil Taxonomy.

Work on Soil Taxonomy, the system for classifying soils in the United States, began shortly after World War II. As the system was being developed, a number of approximations, which was the name given to each new version, were circulated to scientists around the world for review. The 7th Approximation was published in 1960, and a formal publication in book form was issued in 1975. But changes did not end there; the system continues to be refined and updated on a regular basis. An important feature of Soil Taxonomy is its distinctive hierarchical structure. At the top, all of the world's soils are placed into broad groups called soil orders, of which there are twelve. Each order is then broken down into increasingly narrow subdivisions: orders › suborders › great groups › subgroups› families › series. This book focuses on the twelve orders because they are a useful category for discussing both soil formation and soil evolution.

Early in the 20th century, it was widely believed that soils could be classified using geology alone. For example, C.F. Shaw proposed in 1927 that California soils could be placed in five groups based on the kind of parent material from which they formed. Systems such as this were tried in other parts of the world as well, including France and Japan, but it soon became clear that soil classifications based on geology or parent material alone were of limited use. Such systems worked only in areas where the change from parent material to soil had been minimal, such as in deserts, very cold areas, or in recent alluvial deposits. Jenny (1941) provides some insight into how the idea that geology largely determines the nature of soil came about and why this concept was soon discarded. "Technical expressions like limestone soils or granitic soils are encountered in the oldest textbooks on agricultural subjects. They clearly convey the importance of parent material in soil formation." Jenny continues, "However, it remained for Hilgard in America and, independently, for Dokuchaiev in Russia to enunciate the important discovery that a given parent material may form different soils depending on environmental conditions, particularly climate and vegetation."

We now know that, on a regional basis, climate and parent material largely determine the nature of the soil below ground and the nature of plant communities above ground. At the local level, relief or topography influences the development of both soils and plant communities by modifying climate, mainly by the partitioning of water and energy on the landscape. Uneven topography breaks down a broad regional climate into a mosaic of microclimates. Time also is important, since some interval must pass before solid rock or newly deposited sediments can be turned into soil, and the initial soil evolves and changes with the passage of more time. This chapter concentrates on two of the soil forming factors — climate and parent material. The next chapter focuses on the remaining three — organisms, relief (or topography), and time.

The framers of Soil Taxonomy took great pains to classify soils on the basis of measurable properties instead of external factors such as climate. They succeeded in this, but only partially, because many of the measurements used in Soil Taxonomy remained, of necessity, direct proxies for external soil-forming factors such as climate and parent material. For example, Andisols form in a variety of climates but largely reflect the kinds of parent material from which they formed. In contrast, two other soil orders, Gelisols and Aridisols, form from a variety of parent materials but are restricted to certain extreme climates. For Gelisols, it is a very cold climate and for Aridisols, it is a very dry climate.

Andisols

Soils classified in the order Andisols must, by definition, result from the weathering of ash, cinders, or other materials ejected from volcanoes; but taxonomy does not say this directly. Instead, Andisols are characterized as having andic soil properties, which are described as "unique soil properties associated with materials that are rich in volcanic glass or poorly crystalline minerals." Upon weathering, volcanic ash forms a glassy, colloidal material that is very light in weight and has a very high water holding capacity. It also has the capacity to fix large amounts of phosphorus, making it unavailable to plants. In Soil Taxonomy, these unusual physical and chemical characteristics are defined as "andic soil properties" and are the basis for placing soils in the order Andisols. The soil factor equation for Andisols can be greatly simplified: Andisols = f (volcanic parent material).

Worldwide, about 1,500 volcanoes have been active in the past 10,000 years. Most occur around the so-called Pacific Ring of Fire, which includes such countries as Japan, New Zealand, Indonesia, Chile, Ecuador, and Mexico. America is part of this ring, ranking third behind Indonesia and Japan in the number of active volcanoes, most of which are in the Pacific Northwest, Alaska and Hawaii. Not surprisingly, Andisols are well represented in these parts of the United States, as evidenced by the fact that the official state soils of Washington (Tokul), Idaho (Threebear), and Hawaii (Hilo) are all Andisols.

Vertisols

Vertisols (from Latin *verto*, "turn") are clay-rich soils that shrink and swell with changes in moisture content. During dry periods, the soil shrinks, causing deep, wide cracks to form; upon becoming wet again, the soil expands and the cracks close. Such shrink/swell cycles can pose serious engineering problems and generally prevent the formation of well-developed soil horizons. These extreme changes occur because Vertisols are very high in clay, and much of this clay is of the 2:1 type that is highly expansive upon wetting.

In the United States, Vertisols occur from southern Texas to northern Minnesota and then into Canada; east to west, they range from near the Alabama–Georgia border all the way to California and Oregon. This shows that these distinctive soils can form in a wide variety of climates. Vertisols are commonly associated with grasslands or savannas; and Houston Black, the state soil of Texas, falls well within this concept. The official series description characterizes the native vegetation of Houston Black as being "tall and mid grass prairies."

But there are Vertisols in nearby states that form under a totally different kind of vegetation. Two of these soil series, with the very evocative names of Alligator and Sharkey, occupy millions of acres on floodplains along the Mississippi River and its tributaries. According to official series descriptions, the native vegetation for both the Alligator and Sharkey series is bottomland hardwoods dominated by such species as bald cypress, ash, tupelo gum, swamp maple, oaks, and sweetgum. This is quite a contrast. In Texas, Vertisols are associated with tall to mid grass prairies, while just to the east in Louisiana and Mississippi, the same kind of soil occurs on the floodplains and back swamps of large rivers, beneath dense stands of bottomland hardwoods.

From observations in the United States alone, it is apparent that Vertisols can form under a wide range of climates and beneath contrasting kinds of vegetation, which brings us to the subject of parent material. Vertisols occur in many parts of the world, and commonly form from basalt, an igneous rock that is rich in magnesium and iron and low in silicon. Because of its chemical composition, basalt typically weathers to form very clayey soils in which clays of the expansive 2:1 type often predominate. For example, large areas of Vertisols in Africa and on the Indian subcontinent formed from basalt. But a significant number of Vertisols are derived from other kinds of parent material. For example, the official description of the Houston Black soil series informs us that it formed in "clayey residuum derived from calcareous mudstone." What about Alligator and Sharkey? Neither of these soils developed from basalt; instead, both formed from "clayey alluvium of the Mississippi floodplain and its tributaries."

Basalt, mudstones, clayey alluvium — these do not seem to be the same parent material or even to be similar parent materials. But in one important essential they are very similar. Either they are rock that weathers to form a very clayey residuum or they consist of very fine particles deposited by slowly moving or still water, as is common in abandoned channels and low terraces of large meandering rivers such as the Mississippi.

Histosols

Histosols are the third order in which soils are largely defined by the nature of the parent material from which they form. By definition, they are derived from dead plant tissue, generally under waterlogged conditions in swamps and bogs. Not surprisingly, the resulting soils consist primarily of highly decomposed or fibrous organic matter. The name of the order comes from the Greek word *histos*, meaning tissue. In the United States, large concentrations of these soils occur in recently glaciated areas of Minnesota, Wisconsin, and Michigan. In the South, Histosols are especially abundant in wet, low-lying areas of eastern North Carolina, southern Louisiana, and Florida.

Histosols, Andisols and Vertisols are the only soil orders categorized almost exclusively on the basis of parent material, but the impact of parent material is evident throughout Soil Taxonomy. When assessing the role of parent material in soil formation, the two most important things to consider are ease of weathering and richness of weathering. Ease of weathering refers to how quickly parent material can be broken down. Chemical rock weathering is a wet chemistry process, so it is expedited when rain or water from melted snow moves down into the ground instead of running off the surface. This occurs most readily if the surface rocks have already been ground up or pulverized. Fortunately, this has happened over much of the world's land surface. Large parts of the United States, for example, are covered with finely-divided glacial till, loess, coastal plain sediments or river alluvium, and these materials are much more easily weathered than hard intact rock.

Richness of weathering, like ease of weathering, is a fairly straight-forward concept. Rocks or sediments that are low in silicon and high in elements such as magnesium, calcium and iron are said to have a high richness of weathering. They sometimes are referred to as mafic or basic, with basalt being a common example. In contrast, rocks or sediments that are high in silicon and low in these elements are referred to as acidic and are said to have a low richness of weathering. Granite is the most common example. Table 7.1 shows some of the differences between granite and basalt when it comes to elemental composition and the relative richness of weathering. It is clear that granite is much higher in silicon, potassium and sodium than basalt, while basalt is much higher in iron, calcium and magnesium.

Table 7.1. Average elemental composition (percent by weight) of some granite and basalt samples

From Clarke (1924) and Harrison (1933)

Element	Basalt	Granite
Silicon	23.6	32.9
Iron	6.1	2.3
Calcium	8.8	1.3
Magnesium	4.7	0.4
Potassium	1.5	3.4
Sodium	1.7	2.7

Climate and Soil Formation

In deserts and in very cold areas, soils typically reflect the underlying geology, but as rainfall and temperature increase, the nature of the original rock becomes less important. Soils throughout the world are highly correlated with climatic factors such as temperature and the amount and seasonal distribution of rainfall; and this is reflected in modern soil classification systems. Both America and Russia span entire continents and, as a result, have many contrasting areas of climate and natural vegetation. A little over a century ago, scientists in both countries began to classify and map their nations' soils, and not surprisingly, their classification systems relied heavily on climate.

Gelisols occur in climates so cold that parts of the soil remain frozen throughout the year. The name comes from the Latin *gelare*, to freeze. For a soil to be classified in the order Gelisols, it must have permafrost (a frozen layer) or gelic materials (evidence of frost churning) within a meter of the surface. Gelisols are extensive in Russia, Canada, Alaska, Antarctica, and the extreme southern part of South America. Tanana, the state soil of Alaska, is a Gelisol. Conceptually, the soil factor equation for Gelisols is a simple one; Gelisols = ∫ (a very cold climate).

Aridisols, from the Latin *aridus* (dry), are the dry or very salty soils of deserts. In Soil Taxonomy terms, a soil cannot be classified as an Aridisol

unless it has an aridic moisture regime or a salic horizon. An aridic moisture regime means that the soil is dry for much of the year and cannot be
used to grow crops without irrigation. Wetter desert soils, often in low-
lying areas, can be classified as Aridisols if they have a salic horizon, which
is a layer very high in salt. Because of the high salt levels, these soils are
physiologically dry, which means that most plants have trouble extracting
water from them. Deserts occupy a large part of the western United States,
and as a result, so do Aridisols. The state soils of Arizona, Nevada, New
Mexico, Utah, and Wyoming are all Aridisols. The soil factor equation for
Aridisols can be greatly simplified; Aridisols = f (a very dry climate).

Pioneering Studies

Early in the 20th century, Hans Jenny and his colleagues conducted
studies in different parts of the United States to explore how rainfall and
temperature affect the amounts of clay and organic matter in soils. They
began these studies in the 1930s, showing that scientists at that time had
already recognized the critical roles that these two constituents play in the
soil. As discussed earlier, clay and organic matter together are responsible
for the electrical nature of soil, and organic matter is the basis of nearly all
of the biological activity in soil. Jenny concentrated his studies in two areas
of the United States. One was a broad, uniform loess belt that extends
across Kansas and into Missouri. Jenny wrote (1941).

> Here it is relatively easy to select large areas that have great varia-
> tions in rainfall but little variation in annual temperature. The exten-
> sive loess mantle...yields a parent material of remarkable homogeneity...
> In these regions, topography is level, undulating, or slightly rolling. By
> restricting inspection to level areas or to the crests of the loess eleva-
> tions, the factor topography also is kept constant.

Jenny used this area to assess the effects of increasing rainfall on various
soil properties.

In one study, soil samples were taken along a transect running west to
east across Kansas and into Missouri. Mean annual temperature remained
constant along this transect at 11°C (51.8°F), while annual rainfall ranged
from less than 20 inches (51 centimeters) to more than 40 inches (about a
meter). The content of organic matter in the soil surface increased steadily
with increasing rainfall, resulting in a correlation coefficient (r) between
rainfall and organic matter content of 0.946 (1.00 would be a perfect
straight line).

Jenny also studied the effects of temperature on soil organic matter, using samples from a transect running from Louisiana to Canada. From south to north, mean annual temperature decreased from 70°F to 32°F; and as average temperature decreased, the content of organic matter in the soil surface increased from around two percent in the southern United States to around ten percent by weight at the end of the transect in Canada. Here is how Jenny (1941) summarized the effect of temperature on soil organic matter: "The following empirical rule applies: for each fall of 10°C in annual temperature...organic matter content of the surface soil increases from two to three times."

Another study area chosen by Jenny and his colleagues was the Piedmont, a broad belt of igneous and metamorphic rock that extends north to south along the east coast of the United States. Here is how Jenny (1941) described the area.

> In the Eastern part of the United States an extensive region of igneous and metamorphic rocks extends from Maine to Alabama. For the past 40 years, extensive studies of this area by soil scientists of the U.S. Department of Agriculture have resulted in...over 1,000 mechanical analyses of a great variety of soils...these data are admirably suited for a quantitative study of the effect of temperature on the clay content of the soil. Unlike sedimentary deposits, the igneous rocks do not contain clay particles; hence the clay content...is the result of weathering and thus furnishes an index of the intensity of rock decay [and of clay formation].

Jenny's Piedmont studies showed that the clay content of soil increases in a predictable fashion as the climate becomes warmer. One comparison was made between soils derived from igneous rocks on the northern Piedmont (New Jersey and Maryland) and soils derived from similar rocks on the southern Piedmont (North Carolina, South Carolina and Georgia). The average annual temperature in the south was 60.4°F, compared with 54.0°F in the north. The average clay content of the Southern samples was 40 percent by weight, more than twice that of the Northern samples, which contained an average of 19 percent clay. The linear correlation between temperature and clay content for these data was highly significant, with a correlation coefficient (r) of 0.81. The increase in clay with increasing temperature was in basic agreement with van't Hoff's rule, which states that for every 10°C (18°F) rise in temperature, the rate of a chemical reaction increases by a factor of 2 or 3.

In addition to increasing with higher temperatures, the amount of clay in soil increases with increasing rainfall. Jenny and Leonard (1934) demonstrated this by once again using a west to east transect through the Kansas and Missouri loess belt along which rainfall steadily increased. As annual rainfall increased by ten inches (25 centimeters), clay content of the soil surface increased from about 20 percent to nearly 30 percent by weight. The linear correlation between rainfall and clay content was highly significant, with a correlation coefficient (r) of 0.82. These early studies allow us to propose some general rules concerning the effects of climate on soil organic matter and clay.

1. At a given moisture content, the amount of clay in soil increases as the climate gets warmer.

2. At a given temperature, the amount of clay in soil increases as the climate gets wetter.

3. At a given moisture content, the amount of organic matter in the soil surface decreases as the climate gets warmer.

4. At a given temperature, the amount of organic matter in the soil surface increases as the climate becomes wetter.

In a broad sense, the amounts of clay and organic matter in soils are highly correlated with annual rainfall, but such correlations become much better if one considers not only the total amount of rainfall but also how it is distributed throughout the year. For example, does the same amount fall each month, or does more rain fall during the summer, or during the winter? The amount of this rain that evaporates from the surface or is taken up and returned to the atmosphere by growing plants also is important. In addition to considering how much rain falls each year and when it falls, one must consider the amount that is returned to the atmosphere through evaporation and transpiration, usually combined in the catchall term evapotranspiration. When it comes to soil formation, the balance between precipitation and evapotranspiration is much more relevant than precipitation alone. Just as temperature determines the intensity of weathering, the balance between precipitation and evapotranspiration determines the intensity of leaching; and these two parameters, intensity of weathering and intensity of leaching, largely determine the direction and end result of soil formation. We will explore these concepts by using some specific examples, starting with an area in which the intensity of leaching is very low.

Aridisols

The area around Tucson, Arizona, is a typical desert, with sparse vegetation dominated by cacti, desert shrubs, and other drought-tolerant plants. According to the National Weather Service, the area gets less than 12 inches (30 centimeters) of rain annually, while potential evapotranspiration is many times that. Scientists from the Western Regional Climate Center measured the amount of water lost through evaporation each month at a site near Tucson for more than 20 years (1982 to 2005), and the numbers were rather impressive. But before discussing the data and what they mean, it is important to consider how such measurements are made in the United States.

The Oxford Dictionary defines evapotranspiration as "the process by which water is transferred from the land to the atmosphere by evaporation from the soil and other surfaces and by transpiration from plants." The concept is not hard to understand, but actually measuring it is extremely difficult and error prone. Because of this, evapotranspiration usually is estimated using measurements of evaporation from a free water surface (pan evaporation) or by using climatic models. The National Weather Service Class-A evaporation pan is a round metal container ten inches deep and four feet in diameter (25 centimeters by 1.2 meters). The pan is placed in an open area, carefully leveled, and then filled with water to a depth of eight inches (20 centimeters). The water level is measured daily and more is added anytime the level drops by more than an inch. Daily measurements are adjusted as needed for rainfall. Pan evaporation is always greater than actual evapotranspiration from nearby land areas. This is because all of the water in a pan eventually evaporates, but a certain percentage of rain falling on the land will sink into the ground or be held indefinitely by the soil. On average, the continental United States receives about 30 inches (76 centimeters) of rainfall annually; about 20 inches (51 centimeters) of this are returned to the atmosphere through evaporation and transpiration, while the remaining 10 inches (25 centimeters) sink down into the soil, where much of it eventually makes its way to streams.

Measurements of total evaporation usually are adjusted using a pan coefficient or Kp to arrive at a reference value for evapotranspiration. A widely accepted Kp value is 0.70 (Farnsworth and Thompson 1982). Assume that, in a given month, six inches of water evaporate from an open pan. Multiplying this value by 0.70 gives a reference evapotranspiration amount for the month of 4.2 inches (6.0 x 0.70). Unless indicated otherwise, evapotranspiration values in this book are pan evaporation amounts multiplied by 0.70. This, of course, yields only an estimate of true evapotranspiration, but real values are nearly impossible to get under most conditions. Estimates based on pan evaporation, as derived here, are rather crude and imprecise, but they are sufficient for the kinds of broad regional comparisons being made. Because evapotranspiration is, of necessity, such an imprecise measurement, values for rainfall and evapotranspiration in this book are rounded to the nearest whole number.

Table 7.2 shows rainfall and evapotranspiration summaries for Tucson, Arizona. Rainfall data are from the National Weather Service and evaporation data (1982–2005) are from the Western Regional Climate Center. Data are presented for two seasons, the warm-to-hot half of the year and the cool-to-cold half of the year. The warm-to-hot half (April through September) consists of late spring, summer and early fall, approximating the time of year during which the Northern Hemisphere is tilted toward the sun. This is the period of maximum photosynthesis and maximum evapotranspiration. The cool-to-cold half of the year (October through March) consists of late fall, winter and early spring. This approximates the time of year during which the Northern Hemisphere is tilted away from the sun, a period during which most plants are dormant and evapotranspiration is at a minimum.

Table 7.2. Seasonal rainfall and evapotranspiration (inches) at Tucson, Arizon

	Rainfall	Evapo- transpiration	P-E
October–March	6	23	-17
April–September	6	55	-49

Evapotranspiration is much higher during the warm to hot half of the year, but it really does not matter that much in the hot desert. Total rainfall is so low that evapotranspiration greatly exceeds rainfall throughout the year. Note the large negative numbers for P-E (precipitation — evapotranspiration) in the last column of the table. On average, Tucson receives about one inch of rain monthly, but evapotranspiration can return more than six inches (15 centimeters) of water to the atmosphere each month. Very low leaching intensity is the key feature of areas in which Aridisols predominate, and this includes much of the American West. As already noted, the official State Soils of Arizona, New Mexico, Nevada, Utah, and Wyoming are all Aridisols.

We now have introduced five of the twelve orders in Soil Taxonomy and the rationale for recognizing each of them is quite straightforward.

- Aridisols — the very dry soils of deserts
- Gelisols — the very cold soils of circumpolar regions and high elevations
- Andisols — soils that form from volcanic material
- Vertisols — very clayey, high shrink/swell soils
- Histosols — organic soils

The remaining soil orders have more complicated stories, and when it comes to studying them, the continental United States is an ideal natural laboratory. If you key a phrase such as "soil orders map USA" into a search engine, you can quickly access a map showing the geographic distribution of soil orders across the United States. Make sure this map is based on soil survey data from the Natural Resources Conservation Service (NRCS). Such maps are built from the "ground up" and thus are very accurate. A soil order map of the United States shows that three of the orders are well-represented in the West: Gelisols in Alaska, Andisols in the Pacific Northwest, and Aridisols in the Western deserts. East of the Rocky Mountains, the map shows four distinctive soil regions that largely reflect different combinations of parent material and climate:

- Mollisols of the Great Plains
- Alfisols of the Midwestern states south of the Great Lakes
- Spodosols of northeastern New York and northern New England
- Ultisols of the southeastern United States

Other soils occur in each of these regions, but Mollisols, Alfisols, Spodosols and Ultisols represent what soil scientists of the last century often referred to as the "zonal soils." In each zone or region, they are the dominant soils on stable upland areas where the factors of soil formation are best able to leave their distinctive imprints on the landscape.

Mollisols

Imagine you have been transported back in time and are standing somewhere in Nebraska or Iowa a thousand years ago. Looking around, you see no signs of modern life, no highways, no buildings, no planes flying overhead, and no checkerboard pattern of tilled fields. What you do observe is a seemingly unending sea of grass, shoulder high or even head high to an adult human. But you are unable to see the 70 percent or more of the biomass hidden beneath the surface, consisting mostly of fibrous grass roots. About a third or more of these roots die off each year and are replaced, resulting in steady additions of new organic matter to the soil. Over time, this creates the thick, dark surface layer that soil taxonomists now refer to as a mollic epipedon — which is the main feature used to place soils in the order Mollisols. The State Soils of Nebraska, Illinois, Iowa, Kansas, Montana, North Dakota, Oklahoma, and South Dakota are all Mollisols, indicating that these soils clearly dominate the former grasslands of America's interior. These are some of the most fertile and productive soils in the world, partly because leaching intensity is so low that soil nutrients remain in the soil instead of being carried away to nearby streams. In addition, the nutrients in these soils are constantly renewed, as explained by Keller (1957).

> The soils are replenished continually by nutrient reserves contained in rock and mineral particles blown onto the soil by the wind from Missouri River flood deposits and from the western dry plains. This central region of the United States therefore has been provided with an extraordinarily full measure of geologic materials on which chemical weathering can produce a superior soil.

In addition to rich parent material, these Mollisols have a very favorable climate, with most of the annual rain falling during the growing season. Table 7.3 shows climatic summaries for Lincoln, Nebraska, which is in the heart of the Mollisol region. The data are from the website Southeast Lincoln Weather (gwwilkins.org). Note: Precipitation values in this book are given as rainfall equivalents, although we know that in places such as Nebraska much of the precipitation during the winter is in the form of snow. Meteorologists routinely convert snowfall to equivalent amounts of rain, a conversion that depends mostly on temperature. As temperature drops and the snow becomes looser and fluffier in texture, increasing amounts of snow are required to equal the same amount of rain.

During the period of record, Lincoln received nearly 32 inches (81 centimeters) of rain yearly, and about 25 inches of this, nearly 80 percent of the total, fell during the warm to hot half of the year, April through September. Interestingly, this rainfall amount was matched by a nearly equal amount of evapotranspiration, about 24 inches, so there was little leaching.

Table 7.3. Seasonal rainfall and evapotranspiration (inches) at Lincoln, Nebraska (2010–2017)

	Rainfall	Evapo-transpiration	P-E
October–March	7	7	0
April–September	25	24	1

Rainfall was much lower during the cool and cold months, but so was evapotranspiration, so that evapotranspiration and rainfall were in equilibrium. The fact that rainfall and evapotranspiration are in balance throughout the year shows that this is an area of low leaching intensity. Other weather stations in the Great Plains, from North Dakota to Oklahoma, show a similar pattern. As a result of this climatic balance, the nutrient-rich parent materials of the region are subject to only a moderate degree of leaching, so that significant amounts of calcium and magnesium are retained, as well as much of the silicon. With an abundance of these elements, it should surprise no one that the fertile soils dominating this region are rich in 2:1 clays such as montmorillonite.

Alfisols

Once again assume that you have time-traveled to Nebraska or Iowa a thousand years ago, but this time you start walking east, continuing your journey for many days. In time the seemingly endless grasslands start giving way to scattered woodlands, and eventually the transition from grassland to forest is complete. Standing almost anywhere in what is now Indiana or Ohio, you would be surrounded by towering oaks, maples, hickories, black walnut, yellow poplar, and many other tree species. Instead of Mollisols, you would find that most of the upland soils are Alfisols. Miamian, the state soil of Ohio, is an Alfisol, as are the state soils of nearby states such as Indiana, Kentucky, Minnesota, New York, and Wisconsin. This is an indication of just how common Alfisols are in the formerly glaciated areas of America bordering the Great Lakes. Alfisols occur in other parts of the United States, but nowhere else do they occur in such a large contiguous area. The parent material of these Alfisols is similar to that found in the Mollisol region — young, base-rich glacial deposits. The main difference between the Mollisol region and the Alfisol region when it comes to soil formation seems to be climate. Table 7.4 shows average rainfall and evapotranspiration by season for Cleveland, Ohio. Rainfall amounts are from the

National Weather Service and evapotranspiration data are modeled values from the Northeast Regional Climate Center (DeGaetano 1994).

Table 7.4. Seasonal rainfall and evapotranspiration (inches) at Cleveland, Ohio

	Rainfall	Evapo-transpiration	P-E
October–March	18	4	14
April–September	21	20	1

Consider how Cleveland's climate differs from that of Lincoln, more than 800 miles (1,287 kilometers) to the west. Average annual temperature is nearly the same for the two locations at about 52° F. However, it is a different matter when it comes to rainfall. North central Ohio receives more rain each year than eastern Nebraska, about 39 inches versus 32 inches (100 versus 80 centimeters), and nearly all of that difference is in the colder months. Rainfall amounts are similar from April through September, but during the cool to cold half of the year, Cleveland receives more than twice as much rain as Lincoln.

In eastern Nebraska there is a close balance between rainfall and evapotranspiration in both summer and winter, creating an environment of low leaching intensity in which there is little loss of ions such as Ca^{2+}, K^+ and Mg^{2+}. As a result, the clays and organic matter in the Mollisols that form there can retain a lot of these basic ions on their charged surfaces. According to Soil Taxonomy, Mollisols must have a base saturation of 50 percent or more throughout most of the soil profile. This means that more than half of the negative charges on clay and organic matter surfaces are occupied by ions such as Ca^{2+}, K^+, Mg^{2+} and Na^+, while less than half are occupied by H^+ or Al^{3+}. This typically requires a climate of low leaching intensity acting on a base-rich parent material, a condition that is met throughout most of the Mollisol region.

Farther to the east, in areas such as Indiana and Ohio, where Alfisols are predominant, soil profiles undergo significantly more leaching during the cool to cold half of the year. Mostly as a result of this higher winter rainfall, bases are leached from the landscape more thoroughly in northern Ohio than in eastern Nebraska; and the soils reflect this. Instead of Mollisols, the stable uplands in the formerly glaciated parts of Ohio and nearby states are dominated by Alfisols. Despite the leaching effects of winter rains, base-rich parent materials make it possible for Midwestern Alfisols to retain relatively large amounts of bases such as Ca and Mg on their clay and organic matter surfaces. This helps to make them very productive for agriculture. Together, the broad expanse of Mollisols and Alfisols that occupy America's heartland comprise what is often referred to as the breadbasket, so called because this region accounts for so much of America's grain production.

Spodosols

Honeoye, the state soil of New York, is an Alfisol. The word "Honeoye" is taken from an old Iroquois legend that tells of a young man who was bitten on the finger by a snake. He cut off the finger to get rid of the poison and later described where the event had occurred; this is the origin of the name Honeoye, or the place "where the finger lies." Honeoye soils are part of a broad band of Alfisols beginning south of Lake Ontario in New York and then running across the middle of the state from west to east. This belt contains some of New York's best agricultural land and, along with a smaller belt along the New York–Vermont border, is the last major outpost of Alfisols in the northeastern United States.

When we compared Mollisols of the Great Plains to Alfisols of the formerly glaciated areas south of the Great Lakes, we focused largely on climate. The parent materials of the two regions are similar, but the Alfisol region gets more rain, particularly during the winter. Of course, vegetation also is different; the Mollisols developed primarily under grass, whereas the Alfisols formed primarily under forest vegetation. We will consider the important issue of grass versus forest vegetation in the next chapter. But for now, let us focus on the rather abrupt transition from Alfisols to Spodosols that takes place roughly in the middle of New York State.

New York Alfisols formed mostly in base-rich glacial materials. In contrast, the Spodosols to the north and east formed from silicon-rich rocks such as granite on cold, rainy landscapes of the Appalachian highlands. The state soil of Maine is a Spodosol, as are the state soils of Vermont and New Hampshire. These Spodosols formed from parent materials that are much more acid and poorer in bases than the limestone-influenced glacial materials that blanket large parts of Midwestern states such as Wisconsin, Ohio, and Indiana. In addition to forming from different kinds of parent materials, Midwestern Alfisols and Northern Spodosols form under different climates. Table 7.5 shows climatic summaries for Bangor, Maine. Rainfall data are from the National Weather Service and evapotranspiration values are from the Northeast Regional Climate Center, based on the MOREC model (DeGaetano 1994).

Table 7.5. Seasonal rainfall and evapotranspiration (inches) at Bangor, Maine

	Rainfall	Evapo-transpiration	P-E
October–March	21	3	18
April–September	21	13	8

Bangor and vicinity get about 42 inches (107 centimeters) of rain annually, with the summer and winter seasons getting about the same amount. The positive numbers in the last column of Table 7.5 indicate that rainfall exceeds evapotranspiration throughout the year. Winter rainfall exceeds evapotranspiration by about 18 inches (46 centimeters). This is because winter temperatures in Maine are so low that evapotranspiration is minimal. But there is more. Summer temperatures also are relatively low in Maine, so rainfall exceeds evapotranspiration during that season also, meaning that leaching occurs throughout the year. A major difference between the climates of Cleveland (representing the Alfisol region) and Bangor (representing the Spodosol region) is the intensity of summer leaching. Both areas are subject to leaching during the winter months, but leaching is minimal in the vicinity of Cleveland during the warm to hot half of the year. In contrast, landscapes near Bangor are subject to strong leaching both in summer and winter. Maine is also colder than Ohio, meaning that the intensity of weathering and the rate of clay formation are much lower there.

Ultisols

A soil order map of the eastern United States shows that Alfisols and Spodosols start giving way to Ultisols as one moves south from areas that were covered by ice during the last glaciation to those areas that remained glacier free. An extensive region dominated by Ultisols then extends south to the Gulf of Mexico and from the eastern margins of Texas and Oklahoma all the way to the Atlantic coast, interrupted only by the broad Mississippi River floodplain. It is instructive that the state soils of Alabama, Delaware, Georgia, Louisiana, Maryland, New Jersey, North Carolina, South Carolina, Tennessee, and West Virginia are all Ultisols. As mentioned earlier, the heartland of America, dominated by Mollisols and Alfisols, is often referred to as the nation's "bread basket." In a similar vein, the southeastern United States is commonly referred to as the nation's "wood basket." This area, while accounting for just two percent of the world's forestland, produces nearly 20 percent of the world's forest products, and most of this production takes place on Ultisols. In 2011, this region accounted for nearly 50 percent of America's saw-log products, more than 60 percent of its veneer products, and about 75 percent of its pulp and paper products (USFS 2014). Soil taxonomists commonly summarize the differences between Ultisols and Alfisols as follows: Alfisols typically form from base-rich parent materials in cool, humid climates under forest vegetation. Ultisols also form under forest vegetation, but from base-poor parent materials under warm, humid climates.

Statements similar to these are common on soils-related websites and in government and university publications dealing with soil. There are some exceptions to this generalization, but it applies very well to the eastern half of the United States. Let us first consider parent material. Southeastern Ultisols form mostly from granitic rocks or from granitic sediments that were eroded from the ancient Appalachians. These parent materials are base-poor and acidic in comparison to the limestone-influenced till and loess from which Midwestern Alfisols typically form. The Alfisol region also differs from the Ultisol region to the south in terms of climate. The Ultisol region is hotter and has a longer growing season, so the soils are deeper and the already base-poor parent material is more intensely weathered. As a result, Ultisols are higher in iron oxides and 1:1 clays such as kaolinite; and because they contain fewer bases and less organic matter, Ultisols are more acidic and less fertile than Midwestern Alfisols.

Rainfall increases from north to south in the eastern United States, so one might think that leaching intensity also is higher in the south; but the data do not support this. Table 7.6 shows seasonal climatic summaries for Cleveland, Ohio and for Athens, Georgia.

Table 7.6. Seasonal rainfall and evapotranspiration (inches) at Cleveland, Ohio, and Athens, Georgia

October–March (cool to cold half of year)

	Rainfall	Evapo-transpiration	P-E
Cleveland	18	4	14
Athens	24	12	12

April–September (warm to hot half of year)

	Rainfall	Evapo-transpiration	P-E
Cleveland	21	20	1
Athens	22	25	-3

Athens gets about six inches (15 centimeters) more rain than Cleveland during the cool to cold half of the year, but evapotranspiration in Athens is about three times greater during that period, more than compensating for the higher rainfall. As a result, winter leaching intensity at the two locations is about the same; if anything, leaching is slightly greater in the vicinity of Cleveland. Athens get about one inch more rain than Cleveland during the warm to hot half of the year (April–September), but evapotranspiration in Athens is five inches greater. As a result, summer leaching might be slightly higher in Cleveland than in Athens. It appears that the increased evapotranspiration rates in Georgia more than compensate for the higher rainfall during both seasons of the year.

Let us now make another comparison, this time between southeastern Ultisols and Spodosols of the northeast. Because of a fortuitous geologic history, many of the Ultisols in places such as Georgia and Alabama formed from the same kind of granitic rocks that give rise to Spodosols in northern New England. This enables us to see how different climates can produce starkly different soils even when acting on similar parent materials. Ultisols on the Georgia and Alabama Piedmont, for example, tend to be very deep to bedrock and to have red, highly oxidized argillic horizons that typically contain 40 to 60 percent clay. In comparison, upland soils in northern New England derived from the same kinds of granitic rock are much shallower, much coarser in texture, and tend to form spodic rather than argillic horizons.

Climate and Soil Formation Summarized

The depth to which rocks weather and how deep the resulting soils are can be predicted quite reliably by mean annual temperature. The effect of temperature on rock weathering and soil formation is in close agreement with van't Hoff's rule, which states that for every 10°C (18°F) rise in temperature, the rate of a chemical reaction increases by a factor of two or three. Studies have shown that both clay formation and the breakdown of organic matter in soils are highly temperature dependent and both follow van't Hoff's rule.

In addition to the intensity of weathering, the direction of soil formation depends to a great extent on the intensity of leaching, which is best
expressed as the balance between precipitation and evapotranspiration. In assessing how this balance affects soil formation, yearly averages
are of limited use. Because of seasonal effects, this book divides the year
into two equal parts when assessing leaching intensity — a warm to hot
season extending from April through September and a cool to cold season
extending from October through March. These dates, of course, apply only
to the Northern Hemisphere; they are reversed in the Southern Hemisphere. Areas away from the equator have distinct winters and summers
because our planet is tilted on its axis at an angle of 23.5°. This came about
more than four billion years ago when Earth was struck by a Mars-size
planetary body called Theia. This massive collision created our moon and
knocked the Earth permanently off its axis.

We have seen that winter leaching is associated with the formation of
forest soils such as Ultisols and Spodosols, while the Mollisols that dominate much of America's heartland occupy an environment in which precipitation and evapotranspiration are roughly balanced both in summer and
winter. As a result, leaching is limited. Looking at precipitation and evapotranspiration on a seasonal basis enables us to make some useful general-
izations as to why certain soil orders are associated with certain regions of
the United States. For example,

- Aridisols occupy the Western deserts, areas where evapotranspiration greatly exceeds rainfall throughout the year. As a result, overall
leaching intensity is very low.

- A broad expanse of Mollisols occupies the former grasslands of the
Great Plains, an area endowed with young, base rich parent materials.
Most of the annual precipitation falls during the warmer months
when evapotranspiration rates also are high; only a small amount of
rain falls during the cold season. As a result, leaching is limited both
in summer and winter.

- In the eastern half of the United States, both Alfisols and Ultisols
form under forest vegetation in areas where leaching occurs during
the winter months, but where it is minimal during the summer
months. Alfisols develop mostly from young, base-rich parent materials, while southeastern Ultisols develop primarily from base-poor
parent materials derived from granitic rocks.

- The Spodosols of northeastern New York and northern New England
form mostly from base-poor parent materials such as granitic rocks

or from glacial debris derived from them. Because of cold temperatures and abundant rainfall throughout the year, these soils are highly leached, with precipitation exceeding evapotranspiration in both summer and winter.

The Argillic Horizon

Argillic or clay-enriched horizons form in soil when clay particles are brought into suspension by water moving down through the upper part of the profile and then redeposited where downward movement of the water ceases. Of course, one or only a few such episodes are not sufficient for an argillic horizon to form. Instead, the process must occur repeatedly, usually over many centuries. Soil scientists have long believed that the argillic horizon is a key indicator of soil genesis, so they have spent a lot of time and effort trying to determine exactly how it forms. Any explanation must account for the fact that argillic horizons are common in Aridisols, Mollisols, Ultisols and Alfisols — a highly diverse collection of soils that occur under a variety of vegetation types and in highly contrasting climates, including dry deserts, semi-arid grasslands and humid forests. How can the same genetic soil horizon form in such contrasting environments? Researchers began studying argillic horizons in the 1950s and by 1975 the fundamentals had been worked out. This research pointed to one common climatic factor that seemed necessary for clay translocation to take place and for argillic horizons to form.

McCaleb (1959), working with Ultisols, determined that the movement of clay from the A horizon to the B horizon is largely a physical process. He theorized that fluctuating moisture levels are responsible for both the movement and re-deposition of clay particles. Similar relationships were observed in other areas. Thorpe et al. (1959), studying Alfisols in the Midwest, noted that clay was brought into suspension by percolating water as individual particles, which then were redeposited where the downward movement of water stopped. Other research also suggested that clay translocation is a straightforward process in which particles are simply mobilized and then moved down physically while suspended in water (Buol and Hole 1961; Khalifa and Buol 1968). But an especially intriguing inference can be drawn from two of these early studies. McCaleb (1959) and Thorpe et al. (1959) noted that clay movement slowed and then stopped as the soil became wet. This suggests that moisture already present in the soil somehow inhibits clay mobilization by any additional water moving down through the profile.

There is more evidence suggesting a dual role for water as a factor both in bringing about clay mobilization and in impeding it. Losche (1967), in the mountains of Virginia, found more clay and more pronounced argillic horizons on dry south-facing slopes than on moist north-facing slopes. Similar north–south relationships have been reported in Michigan (Cooper 1960) and Russia (Shul'gin 1957). The role of drying on subsequent clay movement also can be inferred by observing regional differences. There is little evidence of clay translocation in regions where water percolates through the soil during the whole year. For example, few argillic horizons are found in the high, wet areas of the Appalachian Plateau or British Highlands. However, they are common on adjacent lowlands where the soil becomes partially or thoroughly dry at some period during the year. Information gathered in studies such as these was used to formulate an overall theory of clay translocation in soil. This theory is clearly expressed in Soil Taxonomy (1975), where the movement of clay particles from one horizon to another is characterized as follows, "Wetting a dry soil seems to lead to disruption of the fabric and to dispersion of clay. Once dispersed, the clay is believed to move with the percolating water and to stop where the percolating water stops."

This simple concept helps us to understand something that otherwise might be quite mysterious. How is it that clay movement and argillic horizon formation take place in soils and in climates that are seemingly so different? Argillic horizons are common in Aridisols, Mollisols, Alfisols and Ultisols. These four soil orders occur under a wide range of climatic conditions, especially those related to the amount and seasonal distribution of rainfall. But areas in which argillic horizons form have at least one climatic factor in common. The amount of rainfall and its seasonal distribution may vary widely, but evapotranspiration equals or exceeds rainfall for at least part of the year. Table 7.7 presents rainfall and evapotranspiration summaries for locations representing three regions of the United States in which either Mollisols, Alfisols or Ultisols are considered the zonal soils. Data are presented only for the warm to hot half of the year, April through September, the period during which evapotranspiration is at a maximum.

At all three locations, evapotranspiration approximately equals rainfall during the warm to hot half of the year. Under these conditions, the upper part of most soil profiles will dry out at intervals and then be rewetted with the next rain, thus setting the stage for clay translocation. When the soil is moist, clay particles are held so tightly by the surfaces of soil pores by water films that they are not easily mobilized by additional water moving down through the profile. But as the soil begins to dry, the moisture films holding clay particles in place become progressively thinner. As the soil becomes even drier, there are fewer and fewer moisture films binding particles together, so clay become more subject to mobilization by downward moving water.

Table 7.7. Rainfall and evapotranspiration (inches) at three locations during the season of maximum evapotranspiration, April through September

Location	Rainfall	Evapo-transpiration	P-E
Lincoln, NE (Mollisols)	25	24	1
Cleveland, OH (Alfisols)	21	20	1
Athens, GA (Ultisols)	22	25	-3

Since the removal of clay from a soil horizon depends on the periodic drying out of that horizon, most clay translocation will occur during the time of year when periodic moisture deficiencies exist. In much of the United States, this situation prevails in the summer months. During this season, rain water tends to move some distance into the profile, mobilizing clay particles on the way down. However, with depth, the water is immobilized by small pores or taken up by plant roots, and the clay is redeposited. As this process is repeated year after year, the upper part of the profile becomes progressively coarser in texture, and as a consequence, it becomes even more subject to drying out and the loss of yet more clay. At the same time, the clay content of the illuvial horizon below continues to increase. As the process continues decade after decade, the textural difference between the eluvial and illuvial horizons becomes more pronounced and the boundary between them more abrupt.

Oxisols

Let us now turn our attention to a group of soils that do not have argillic horizons and do not occur in the continental United States. These are the Oxisols, the highly weathered red to yellow soils of the humid tropics and subtropics. Oxisols are very uniform in color and texture throughout their profiles, which commonly are two to three meters deep or more. Boundaries between horizons are very diffuse and hard to identify. The soils commonly are very high in clay content, but the clays are so well aggregated that the soil typically feels like loam or fine sand to the touch; soil scientists use descriptors such as pseudo-silt or pseudo-sand to describe the texture. The clay assemblage is dominated by low activity clays (mostly kaolinite) and large amounts of iron and aluminum oxides.

Oxisols are most common on the ancient continental shield areas of South America and Africa. These are old, highly weathered land surfaces where deep, prolonged weathering and intense leaching have concentrated resistant minerals such as quartz, along with kaolinite and large amounts of iron and aluminum oxides. Outside the continental shields, Oxisols are confined mostly to easily weathered parent materials. For example, Bayamon, the representative soil of Puerto Rico, is an Oxisol, and according to the NRCS website, it is found in areas of limestone.

In addition to very stable landscapes and/or easily weathered parent materials, Oxisols require a tropical climate with abundant rainfall. In order to appreciate the extreme weathering and leaching that are required, let us take a mental journey to the South American country of Guyana and compare the climate there with that of the American state of Georgia. Georgetown, Guyana, and Athens, Georgia, are about 2,500 miles (4,022 kilometers) apart and differ markedly in climate and vegetation; and as one might expect, their soils are different as well. Guyana is located about 5° north of the equator, so it is hot year-around. On average, it is much hotter in Guyana than in Georgia. Note the following north to south temperature progression.

Table 7.8. Mean annual temperature at Bangor, ME; Athens, GA; and Georgetown, Guyana

Location	Mean annual temperature
Bangor, Maine	7°C
Athens, Georgia	17°C
Georgetown, Guyana	27 °C

Athens is about 10°C hotter than Bangor, and as discussed earlier, this is the primary reason why granitic rocks in Georgia produce deep, red, clayey Ultisols, while in Maine, the same kinds of rock are converted to shallow, coarse-textured Spodosols and Inceptisols. Just as Athens is 10°C hotter than Bangor, Georgetown is 10°C hotter than Athens. This means that weathering intensity at Georgetown is two to three times greater than at Athens. This is one reason, but not the only reason, that Oxisols are common in Guyana but not present in Georgia or anywhere else in the continental United States.

High temperatures alone are not sufficient for the formation of Oxisols; another requirement is abundant rainfall and the intense leaching that accompanies it. Table 7.9 presents climatic summaries for Georgetown, Guyana. As discussed above, Guyana is so near the equator that summer and winter do not exist. But Georgetown does have wet and dry seasons, or to be more precise, two brief wet seasons and two brief dry seasons. The "dry" seasons are really not so dry; in fact, the average monthly rainfall (more than four inches or ten centimeters) during the so-called dry seasons would be considered high anywhere but in the humid tropics. The dry months are "dry" only in comparison to the nearly ten inches (25 centimeters) of rain that fall each month during the wet seasons. Rainfall and evapotranspiration data for Georgetown are summarized separately for the seven months of very high rainfall and for the five months of moderate rainfall. Evapotranspiration data are from Persaud (1974) and rainfall data are from the website climatemps.com.

Table 7.9. Seasonal rainfall and evapotranspiration (inches) for Georgetown, Guyana

Season	Rainfall	Evapo-transpiration	P-E
Wet months (7)	68	25	43
Dry months (5)	21	20	1

Annual rainfall at Georgetown is nearly twice that at Athens. In Athens, nearly all leaching takes place during the cold months, while in Georgetown, most leaching takes place during the wet seasons. During the five drier months in Guyana, precipitation and evapotranspiration are about equal, so any rain entering the soil is quickly immobilized or taken up by plant roots before it can move very far down into the profile. Table 7.10 compares rainfall and evapotranspiration during the respective leaching seasons at Athens and at Georgetown.

Table 7.10. Rainfall and evapotranspiration (inches) during the seven wet months at Georgetown, Guyana, and during the six cool to cold months at Athens, Georgia

Location	Rainfall	Evapo-transpiration	P-E
Georgetown, Guyana	68	25	43
Athens, Georgia	24	12	12

During their respective leaching seasons, Georgetown gets nearly three times as much rain as Athens, but the excess of rainfall over evapotranspiration, almost four times greater, is even more striking. In any given year, nearly four times as much water enters the surface and percolates down through the soil at Georgetown than at Athens. The data are clear; both weathering intensity, due to year-around high temperatures, and leaching intensity, due to much higher rainfall, are two to three times greater in Guyana than in the American state of Georgia. Accordingly, Oxisols are common on the stable uplands of Guyana and nearby areas in South America, while Ultisols are the predominant soils on similar landscapes in Georgia and much of the southeastern United States.

But why, in most cases, does the formation of Oxisols require such intense leaching and weathering over such a long period of time? To answer that, let us summarize what happens when rocks are exposed to a tropical climate in which rainfall greatly exceeds evapotranspiration. You will recall that just eight chemical elements (oxygen, silicon, aluminum, iron, calcium, magnesium, potassium and sodium) make up nearly 99 percent of surface rocks by weight. In humid, tropical environments, enormous numbers of tiny, highly active H^+ ions are generated, and they rapidly break down rock structure to release the elements held within. Nearly all of the calcium, magnesium, potassium, and sodium is soon leached away, leaving oxygen, silicon, aluminum, and iron. Most of the iron combines with oxygen to form iron oxides, giving Oxisols their characteristic yellow to red hues. This leaves silicon and aluminum, most of which initially combine to form the 1:1 clay kaolinite.

Kaolinite clay is very stable, but when subjected to year-around high temperatures and high rainfall for centuries, it is slowly but inexorably broken down, releasing silicon and aluminum into the soil solution. Neither of these elements is very soluble, but silicon is the more soluble of the two, and over time, this small difference matters. The soils are slowly but inexorably stripped of silica, leading to a relative concentration of iron and aluminum oxides, which are the hallmarks of Oxisol formation. Some scientists refer to this process as desilication, emphasizing the loss of silicon, while others use the term ferralitization, putting the emphasis on the gradual concentration of iron (ferr) and aluminum (al) in the profile. Whatever you choose to call it, the process of removing most of the silicon from a landscape takes a very long time, even under conditions of intense weathering and leaching. It should not surprise us that, in addition to requiring year-around high temperatures and a lot of leaching rains, Oxisols are mostly found on very old, stable landscapes.

Chapter 4 described how dusting the open ocean with iron caused huge increases in the growth of plankton, increases that soon were followed by impressive increases in salmon populations. These results were possible because the open ocean is very low in dissolved iron, an element that all plants need to grow and reproduce. In areas of the ocean far away from land, lack of iron is the main limitation to biomass production. Ocean water contains 10,800,000 parts per billion of salt, but only 3.4 parts per billion of iron.

It has been shown that iron fertilization increases biomass production in the "iron deserts" of the open ocean. But another group of scientists (Kaspari et al. 2009) investigated the other side of the coin by considering what would happen if salt, NaCl, were used to fertilize old, sodium-depleted land surfaces like the ones in which Oxisols predominate. Choosing a site in the Peruvian jungle more than 1,200 miles (1,900 kilometers) from the ocean, they added NaCl dissolved in water to 35 plots, while adding just plain water to 35 additional plots for comparison. In a short time, the leaf litter in the salted plots began disappearing about 40 percent faster than the controls and the number of termites increased by a factor of seven. One researcher was quoted as saying that, after 18 days, "some of the scariest-looking termites I've ever seen" began swarming over the salted plots (Dell'Amore 2009). The authors concluded that lack of salt can limit the numbers of termites and other decomposers in tropical forests, causing litter to accumulate on the forest floor, sequestering large amounts of carbon. This research suggests that, just as fertilizing the oceans with iron filings might be a way of taking carbon out of the atmosphere, adding salt to some of the world's rainforests could have just the opposite effect by speeding up the decomposition of organic matter.

Indirectly, this underscores the importance of termites (*Syntermes dirus*) in tropical settings, and especially on landscapes in which Oxisols are the predominant soils. Termites and Oxisols have long been linked; wherever you find Oxisols, you are very likely to find termites; and many scientists believe that these busy ground-dwelling insects play a major role in their formation. They argue that termite activity is at least partly responsible for the diffuse horizon boundaries and the very great depths to which many Oxisols extend. In the tropics, termite biomass per unit area equals or exceeds that of earthworms in temperate regions; and they rework the soil in much the same way. In areas of sparse vegetation, these voracious insects might pass thousands of pounds of soil per acre through their bodies in order to get the organic matter they need. In the process, they thoroughly aerate and mix the soil, often to great depths.

Not only do termites ingest a lot of soil, they also ingest a lot of organic matter. On a global basis, about 20 percent of the carbon re-entering the atmosphere each year does so by way of a termite's gut. This is because there are so many termites in the world, an estimated 250 quadrillion, and about one third of each termite's body consists of tiny gut-dwelling bacteria and protozoa that actually process the organic matter. It is estimated that there are more than 30 million termites for every human living on Earth. Considering these numbers, it should be no surprise that termites play such significant roles in the global carbon cycle and in the formation of tropical soils.

As they dig through the ground, termites leave behind voids and tunnels that serve as passageways through which rain water can move rapidly into and through the soil. This increases the effectiveness of leaching and speeds up the removal of silica from the soil profile. Termites travel far below the surface (commonly to ten meters or more) in search of water and clay to use in building their mounds and lining their tunnels. This mining activity deepens the soil and keeps bringing kaolinite to the surface where it is more subject to being broken down into $Al(OH)_3$ and H_4SiO_4. Then, as mentioned above, silicon is preferentially leached from the landscape and the soil profile is increasingly dominated by iron and aluminum oxides. In many cases, so much silicon is removed and the concentration of $Al(OH)_3$ becomes so high that the area can be commercially mined for aluminum ore, or bauxite.

Termites evolved some 200 million years ago, so they have been part of tropical ecosystems for a very long time. When assessing their roles in the formation of Oxisols, it is important to consider not just those that can be observed at present, but also the countless generations that have lived out their lives in some of these very old soils. We should remember that many sites where we currently do not see termite activity probably were inhabited by these busy, soil-eating insects in the past, and the chances are they will come again.

Scientists recently discovered a gigantic complex of termite mounds in a seasonally dry tropical forest of northeast Brazil. A veritable termite megapolis, it covers an area the size of Great Britain (Martin et al. 2018). Dating of organic matter from some of the mounds indicate that they range from 690 to 3,820 years in age. According to a statement issued by the American Association for the Advancement of Science, "The amount of soil excavated...is equivalent to 4,000 Great Pyramids of Giza, and represents one of the biggest structures built by a single insect species."

This discussion of termites is, in some sense, a prelude to the next chapter, which takes a comprehensive look at living organisms and the diverse roles they play in soil formation. For example, the next chapter describes how trees and grasses diverged as life forms and how this divergence is responsible for many of the differences we see between forest soils and grassland soils. In addition to trees and grass, we consider how other organisms alter the soil and other aspects of their environment so much that they have been referred to as "canonical ecosystem engineers." These include termites, earthworms, peat mosses, and not surprisingly, our own species. In addition to organisms, the next chapter examines the importance of topography and time in soil formation.

Chapter 8. Organisms, Topography, and Time

The last chapter described what someone would see if he or she time traveled to eastern Nebraska or western Iowa a thousand years ago and then started walking east. For a few days our time traveler would be surrounded by a seemingly endless expanse of tall grass. But after many days of walking, the grasslands would start giving way to scattered woodlands. Eventually the transition from grassland to forest would be nearly complete, and perhaps somewhere in Indiana our traveler would be surrounded by trees. The change in the dominant vegetation above ground would have been accompanied by equally profound but unseen changes underground. The thick, black surface layers so characteristic of grassland soils would have been replaced by thinner surface layers, and they would increasingly be underlain by light-colored eluvial horizons. As trees became the dominant regional vegetation, Alfisols would replace Mollisols as the dominant regional soils.

Harney, the state soil of Kansas, is a good example of Mollisols, the kind of soil that typically forms under prairie vegetation. A profile picture and a brief description of this soil can be viewed on the NRCS State Soils website. Keying a phrase such as "NRCS state soils" into a search engine will get you to the website. Once there, scroll down to the table labeled Representative and State Soils and click on the factsheet for Harney. A photo of a typical profile is displayed. It shows a dark surface layer nearly two feet (60 centimeters) thick underlain by a lighter colored subsoil (B horizon). Notice that there are no sharp boundaries between horizons; with depth, the very dark surface layer simply grades into the somewhat lighter colored B horizon. Extending over millions of acres, Harney is a classic Mollisol and one of the most productive soils in the world. It once was proposed that Harney be designated America's national soil (Hole and Bidwell 1989), a proposal that is not without merit.

Let us now turn to Bama, the state soil of Alabama, an example of the kind of soil profile that commonly forms under forests in the eastern United States. The profile picture of Bama on the NRCS State Soils website shows three easily discernible horizons. At the surface is a dark A horizon only a few inches thick. This is underlain by a very distinctive light-colored E or eluvial horizon about six inches (15 centimeters) thick. The pale colors of the E horizon are the result of water percolating down through the profile and stripping away clay to expose the pale, often translucent surfaces of sand and silt particles. Clay is removed from the E horizon and then deposited in the bright red argillic (Bt) horizon below. The Bama profile is a good example of the distinctive A-E-B horizon sequence commonly expressed by Ultisols and Alfisols in undisturbed forested areas.

A forest soil profile that is even more striking can be seen in another soil order that forms under forest vegetation, the Spodosols. Turnbridge, the state soil of Vermont, is a good example. Clicking on Vermont brings up the profile picture of Turnbridge. Note the thin but very distinctive surface layer, which in this case is actually a very dark O horizon consisting of relatively undecomposed organic matter. The underlying E horizon, gray to almost white, also is very striking, as is the underlying B horizon, which instead of an argillic, is a spodic horizon. The distinct brown to reddish colors in the B horizon are due mostly to the presence of oxidized iron which was removed from the E horizon by downward moving water and then redeposited below.

Soils of Forests and Grasslands

One of the most important difference between grass and trees relative to soil formation is that grasses are monocots while nearly all trees are dicots. Monocots are so named because they grow from an embryo that produces only one seed leaf or cotyledon, while dicots grow from an embryo that produces two seed leaves (dicotyledons). This rather small difference at the beginning of life leads to profound differences as the plants grow to maturity. There are a number of differences between dicots and monocots, but we will focus on those that are most relevant to soil formation. For example, grasses have dense, fibrous root systems in which all roots are about the same size, much like hair on a person's head. In contrast, trees have a prominent main root analogous to the stem or trunk, with many large, woody lateral roots; and like tree branches above ground, tree roots subdivide repeatedly as they spread through the soil. The root systems of both grasses and trees are a rough mirror image of the above-ground plants. The above-ground parts of a grass plant consist of many individual stems that emerge from near the ground surface, and these stems have leaves but no secondary branching. This is replicated below ground with a root system consisting of many individual roots forming a dense fibrous network, but with little or no branching. The fibrous nature of grass roots is enhanced by the fact that they have many more root hairs than do most other kinds of plants, including trees.

In trees, roots develop from a radicle, which emerges from the bottom of the embryo shortly after germination. But in monocots such as grasses, the radicle aborts soon after germination, so roots arise adventitiously, which means "forming accidentally or in an unusual anatomical position." This certainly applies to grass roots. Emerging from nodes on the lower part of the stem and then growing down into the soil, they are commonly referred to as "prop roots." If you have ever walked through a field of mature corn (*Zea mays*), you probably have seen them, since they are especially prominent in that species.

Another important distinction between trees and grass is that trees are capable of secondary growth while grasses are not. Trees lay down new layers of growth each year and since they sometimes live for centuries, can reach enormous size. The tallest tree measured so far, a California redwood named Hyperion, is 397 feet (121 meters) tall and is 16 feet (4.9 meters) in diameter near its base. In nearly all grasses, the above-ground parts die back each year, so even the tallest species are only a few feet in height, with stems rarely exceeding the thickness of a human finger. Bamboo, of course, is a notable exception to this. Not surprisingly, trees contain a lot of cellulose and lignin, mostly in their trunks but also in large branches and roots. By weight, a mature tree contains about 45 percent cellulose and 25 percent lignin (Robinson 1990; Novaes et al. 2010). In comparison, most grasses contain less than 20 percent cellulose and less than ten percent lignin (van Soest 1982). As a general rule, trees contain more than twice as much lignin and cellulose as grasses do.

In both trees and grass, above-ground growth comes from the meristem. In grasses the meristem is located at the base of the growing plant, at or even below the ground surface. This means that all of the stems and leaves above ground can die back each year without killing the plant. When conditions become favorable, the meristem simply produces new stems and leaves. Having the meristem located at or just below the ground surface enables grasses to survive bitterly cold winters or raging fires, and the above-ground parts can be entirely consumed by grazing animals without killing the grass. In contrast, tree meristems are located at the ends of the branches.

Because of these different life strategies, grasses and trees differ in how much they invest in above-ground growth relative to below-ground growth. According to Harris (1992), "For most trees under normal conditions...the top is 5 to 6 times heavier than the roots." This ratio is reversed in grasses, with grass roots typically weighing two to three times as much as the above-ground plant parts (Sainju et al. 2017). Weaver and Zink (1946) studied several grass species of native Nebraska prairies and found that the roots typically weighed about three times as much as the above-ground biomass. In some species, such as blue grama (*Bouteloua gracilis*), the roots were five times as heavy. The following general rule seems to apply: In forests, two thirds or more of the biomass is above ground, while in grasses two thirds or more of the biomass is below ground.

As a result of the differences outlined above, organic matter turnover is much more rapid in grassland than in forests. At least half of the total biomass in grassland dies back and is replaced each year. This includes nearly all of the above-ground parts and about one third of the root mass. In contrast, only a small percentage of the biomass in forests is recycled each year. Walking through a mature forest and seeing the thick trunks all around you and the massive spreading crowns overhead should convince you of this. Second, about half of the dead plant material in grasslands is deposited underground, mostly in the top six to 12 inches (15 to 30 centimeters) or so of soil; in forest ecosystems, much of the dead plant tissue is deposited on the ground surface in the form of bark, leaves, twigs, cones and, in some cases, entire trees. In addition to being deposited on the surface, dead material from trees tends to be more massive and to contain much more lignin, both of which slow down the rate of decomposition.

Grassland soils typically have thick, dark surface layers in which the organic matter rapidly decomposes and is intimately mixed with the mineral soil particles. To appreciate the end result, readers should once again view the profile picture of Harney, the state soil of Kansas, on the NRCS State Soils website. Alternatively, one could view the profile picture of Tama, the state soil of Iowa or Drummer, the state soil of Illinois, both typical prairie soils. For contrast, readers are encouraged to once again view the profile picture of Turnbridge, the state soil of Vermont, a classic forest soil, on the same website. Turnbridge illustrates the kinds of profiles that typically develop under forest vegetation, where woody, lignified, hard-to-decompose plant materials are deposited mostly on the soil surface.

Silicon and Grass

Another interesting distinction between grass and trees is that grasses concentrate much more silicon in their tissues than trees do. Early studies showed that some species of *Poaceae* (grasses) contained between ten and 20 times the concentration of silicon found in trees (Hodson et al. 2005). Silicon in grasses is believed to serve a number of purposes. Perhaps the most important is structure, with silicon strengthening grass stems in the same way that lignin strengthens the trunks of trees. This enables grass plants to grow very tall relative to their stem diameter without becoming woody. Tall prairie grasses with stems less than two centimeters thick can reach a height of two meters or more, yielding a very high height-to-stem diameter ratio. This is possible largely because the silicon that is preferentially absorbed by grasses makes the stems stronger and more rigid. This is fortuitous, since the biological cost of silicon as a strengthening agent is only five to ten percent that of cellulose and lignin (Raven 1983).

Hodson et al. (2005) summarized data from 125 scientific studies in which researchers determined the silicon concentration of the "leaf and non-woody shoot tissues" for more than 700 plant species. I selected ten common prairie grasses and ten tree species common to the eastern United States from this list and calculated an average silica content. The grasses included such species as big bluestem, hairy grama, and Indiangrass, while the trees included such species as black walnut, American sycamore, red oak, and shagbark hickory. On average, the grasses contained about five times as much silicon per weight as the trees. Two well-known grasses, rice and bamboo, are especially high in silicon. Rice plants commonly contain four to five percent silicon by weight, while silicon can account for as much as ten percent of bamboo plants by weight. High silicon content largely accounts for the great strength of bamboo and its value as a structural material. High silicon content also enhances the capacity of grasses to form dry, decay-resistant hay upon drying out. Because there is so much silicon in grass tissue, the surface layers of many grassland soils are enriched with silica in the form of opal, which is silica combined with varying amounts of water; these opal bodies, commonly called phytoliths, or "plant rocks," can make up 50 percent or more of the surface volume in some grassland soils. We do not as yet know the full implications of silicon concentration in grasses and in certain grassland soils, but it is an interesting phenomenon and one that should be noted for future study.

Earthworms

Because grasslands produce so much organic matter that is recycled so quickly, most grassland soils are teeming with life. The top five to ten inches or so of a grassland soil are home to enormous numbers of grubs, worms, and other invertebrates that feed on grass roots, organic matter, or on other, smaller invertebrates in the soil. For example, a single acre of grassland can contain more than 50 million springtails, a species of wingless insect that lives underground. This same acre may harbor several million earthworms, and even people who know little about earthworms tend to associate them with fertile, productive soils. If someone were to turn over a spade full of earth and find that it contained a lot of earthworms, he or she would assume that the land was good for agriculture, and that assumption would in most cases be correct. In the Netherlands, an area of soil recently reclaimed from the sea failed to develop a good A horizon and remained unproductive. In an effort to correct this problem, Dutch scientists introduced earthworms to the area. Soon after the earthworms had arrived, the soils began to develop dark, organic-rich surfaces and in a short time the crops growing on them began to thrive.

The great naturalist Charles Darwin was an unabashed earthworm admirer. In his last book, The *Formation of Vegetable Mould through the Action of Worms*, published in 1881, he wrote, "It may be doubted whether there are many other animals which have played so important a part in the history of the world as have these lowly, organized creatures." Darwin was the first scientist to study earthworms intensively and write about them, and for some curious reason, many subsequent authors seem compelled to follow in his literary footsteps by, at some point, referring to them as "lowly creatures" as he did. Lowly creatures they may be, but they are admirably suited to life in the soil.

Earthworms do best in soils with a near neutral pH (6.5 to 7.0) and an abundance of well-decomposed organic matter. Since the surface horizons of Mollisols typically have a pH within this range and commonly contain 20 percent or more finely divided organic matter (worm food) by volume, these soils provide an ideal environment in which huge populations of earthworms can thrive. Earthworms also are common in many Alfisols, but they have trouble tolerating the high acidity common to Ultisols and Spodosols, so they are rarer in such soils. They also are less common in soils that are either very clayey or very sandy. Heavy clay is too dense for them to move through with ease and the sharp, angular edges of sand grains can injure their fragile bodies.

In order to appreciate the role of earthworms in mixing and homogenizing the surface layers of soil, it helps to understand a little about their body structure. Earthworms are a member of the phylum *Annelida*, the segmented worms, with bodies consisting of many segments arranged much like a string of pearls. These segments are not fixed in size. If an earthworm needs to pry the soil apart at some point, individual segments can be filled with fluid to expand them. However, if the creature needs to squeeze through a narrow crevice, the appropriate segments can be emptied to make the body very thin. Each segment also has a pair of stiff bristles called setae that grip surrounding surfaces and help propel the earthworm through the soil.

Earthworms spend their entire lives tunneling through the earth and eating huge quantities of soil to get at the organic matter it contains. Some species remain within a few inches of the surface, where they create dense networks of horizontal burrows. They fill these burrows with the nutrient-rich remains of mineral soil and organic matter that have passed through their bodies. In addition to being high in nutrients, this waste material, referred to as casts, has a pH near 7.0. Some earthworms live in vertical burrows that extend from the ground surface to a depth of five feet or more. They travel up to the surface at night to gather organic debris and pull it down into their tunnels. Their presence is easily detected, since they leave little piles of organic debris around the openings to their tunnels. In addition, they deposit most of their excrement on the surface in the form of distinctive casts. The nightcrawler, *Lumbricus terrestris*, commonly used for fishing, is a well-known member of this group.

A single acre of grassland might contain several million earthworms, which is a total biomass equivalent to that of two or more horses. Continuing with the horse analogy, the earthworms in an acre of rich grassland can consume an amount of soil each year equal to the weight of more than 30 horses. Darwin speculated that in some areas the entire surface layer had passed through the bodies of earthworms at least once. He also surmised that these earth-eating ecosystem engineers can create ten or more inches (25 centimeters) of new topsoil in only a century or so. Darwin wrote in 1881, "Farmers in England are well aware that objects of all kinds, left on the surface of pasture-land, after a time disappear, or, as they say, work themselves downward." The objects really were not working their way down into the soil; instead, earthworms were slowly burying them by depositing the digested remains of organic matter and mineral particles on the surface year after year. This is important, because the quality of soil is improved as it passes through the body of an earthworm; it emerges much higher in bases and richer in such nutrients as nitrogen and phosphorus. While adding new material to the surface, earthworms also function as biological rototillers, thoroughly mixing and homogenizing the part of the soil that they inhabit. In time, this creates soil horizons that are uniform in color, texture and consistency throughout.

Though highly regarded by much of the world, earthworms are not admired everywhere. Ecologists in the American state of Minnesota and the Canadian province of Alberta, for example, are waging active campaigns to stop what they perceive as a dangerous earthworm invasion from the south. A University of Minnesota website points out that there were no earthworms in Minnesota prior to the arrival of Europeans because the recent glaciation had wiped them out. Early settlers brought growing plants with them, and the soil in which they grew often contained earthworms or earthworm egg cases. Ships sailing from England and Europe also used rocks and soil as ballast, some of which they later dumped on shore to balance the ship's weight. Earthworms undoubtedly hitched a ride in this material as well. Below is an excerpt from a University of Minnesota website, which presents the case for earthworm exclusion.

> Minnesota's hardwood forests evolved in the absence of earthworms. Without worms, fallen leaves decompose slowly, creating a spongy layer of organic "duff." This duff layer is the natural growing environment for native woodland flowers. But when European earthworms invade a forest, they eat the duff. Big trees survive, but many young seedlings perish, along with many ferns and wildflowers. Some species return after the initial invasion, but others disappear. Once they have invaded, earthworms cannot be removed. The only way to protect our worm-free, flower-filled forests is to prevent new earthworm infestations.

The University of Alberta in Canada has a similar website that raises the same concerns, arguing that "Earthworms can harm other plants and animals when they consume leaf litter from the forest floor." The Alberta website encourages citizens to help by downloading a computer application to track the spread of earthworms. The fact that ecologists in both the United States and Canada are so alarmed about an earthworm invasion of their forests is, in a way, a tribute to the creatures. It acknowledges their status as canonical ecosystem engineers, organisms capable of altering the ecosystem to such an extent that they determine whether other species thrive in the environment or disappear from it.

The North American Beaver

There is more to be written about earthworms, but before continuing that story, let us pay tribute to a species whose role as a soil former and canonical ecosystem engineer is not as well known, the North American beaver or *Castor canadensis*. When Dutch traders bought Manhattan Island from the Lenape tribe in 1626, the island was, according to Goldfarb (2018) "little more than a pot-sweetener." The real prize was the more than 7,000 beaver skins that the Lenape offered as part of the deal. When Europeans first arrived in North America, there were millions of beaver ponds, perhaps as many as 250 million, and some of them were enormous. Goldfarb estimates that the water bodies created by these busy engineers might have covered an area nearly the size of Texas. Beavers, in contrast to earthworms, thrive only in forested areas. They have very strong teeth with a second set of lips behind them, so they can gnaw trees down and drag wood underwater without drowning.

Like all impoundments, beaver ponds slowly fill with sediment, eventually creating a submerged terrain consisting of saturated mineral soil. Shortly after trappers "de-beavered" North America, many thousands of beaver dams were breached and the formerly submerged and saturated sediments were exposed to the air; upon drying out, they became, in Goldfarb's words, "some of the finest soil a farmer could till." Beavers operated as soil formers by modifying topography and the flow of streams, in time creating new landforms and new soils. They were driven nearly to extinction in North America, but they are now making a comeback; and scientists are beginning to recognize their potential to restore degraded ecosystems.

A British Story

Having given the North American beaver his due, let us now return to earthworms, and more specifically to earthworms in the soils of Britain. In 1837, shortly after returning from his famous journey on the *Beagle*, Charles Darwin delivered a paper to the Geological Society of London on the subject of earthworms. He then continued to observe and study them for the next 40 years and published his famous book on worms in 1881.

Darwin grew up in the ancient town of Shrewsbury, on the fertile plain of North Shropshire, which has some of the best farmland in England; and when you think of good farmland in England, you usually are thinking of

Alfisols, which pedologists there have traditionally referred to as brown earths or brown forest soils. For a long time, pedologists in America also referred to Alfisols as brown earths or brown forest soils. These soils occupy nearly half of the land area in England and are responsible for most of the nation's agricultural production. The phrase "brown forest soils" is an apt one, because it reminds us that at one time nearly all of England was forested, and with the kinds of trees that many Americans would recognize. In prehistoric times, about 85 percent of England was covered with hardwood forests very similar to those of the eastern United States. Tree genera such as beech (*Fagus*), oak (*Quercus*), ash (*Fraxinus*), elm (*Ulmus*), and basswood or lime (*Tilia*), for example, are native to the forests of both the eastern United States and England.

There is a difference, however; much of the eastern United States is once again heavily forested, while most of England remains deforested. It is estimated that half of England's forests had been cleared as early as 500 BC. Of course, that is only an estimate, but we have more reliable data from the year 1086, when the Domesday Book was completed. The Domesday Book is the record of the "Great Survey" commissioned by William the Conqueror to count the acres of farmland, the numbers of livestock, the numbers of water-driven mills, in fact, to sum up all of the wealth that he had acquired by invading and conquering England. William's agents found that about 15 percent of his kingdom remained in forest, about 35 percent was cropland (mostly wheat, but other grains as well), and about 30 percent was pasture. The percentage of forest land declined further to about ten percent by 1350, and by the mid-1800s, when Darwin was conducting his studies on earthworms, it had shrunk to below five percent (Rackham 1990).

Darwin was well aware that he was studying ecosystems that had been transformed by humans, and he suspected that the large earthworm populations he observed might, to some degree, be creations of human agriculture. He knew that centuries before his birth, much of England's forests had been replaced by grassland or were planted with wheat, which is a grass. As a consequence, many forest soils had been replaced by grassland soils, which was wonderful news for earthworms. Darwin observed that there were many more earthworms in pastureland than in forests and even described some forested areas in which earthworms were conspicuously absent. In one passage (1881), he describes a soil as being "argillaceous, very poor, and only just converted into pasture, so that it was for some years unfavorable for worms."

In addition to converting many forests to grasslands, the English did something else that enabled earthworms to thrive; they began raising millions of sheep on the deforested lands. Having large numbers of sheep (as well as cattle) on the land was beneficial to earthworms. If earthworms had a religion, it is probable that its high priests would praise sheep and cow manure as manna sent from heaven. Below are some numbers from the extension service at Purdue University (Kladivko 1993) showing the effects that animal manure can have on earthworm populations.

Table 8.1. Effect of land use and treatment on earthworm populations

Land use	Earthworms per acre
Cropland, plowed (corn)	40,500
Cropland, plowed (soybeans)	243,000
Dairy pasture (manured)	1,376,000
Dairy pasture (heavily manured)	5,261,000

The export of wool and, in time, the export of woven cloth were sources of great wealth to England in the Middle Ages and for some years after; and large earthworm populations, by enriching the soil, made these enterprises more productive and more profitable. One can only wonder at how much additional agricultural productivity and national wealth have accrued to many generations of English citizens courtesy of their earthworm partners.

"Very well," the ecologists of Minnesota and Alberta might say, "the English are welcome to them!" The Albertans and Minnesotans have legitimate cause to be alarmed. An organism or life form that is beneficial in one ecosystem can easily be detrimental in another where conditions are different or where other values are at stake. Having walked through some of the beautiful forested areas in northern Minnesota in the spring, I have much sympathy for the concerns of these northern ecologists.

The early inhabitants of England, when cutting down their nation's forests to make way for wheat or for grass to feed sheep and cattle, probably had no idea that they were changing the very nature of the underlying soil. The transformative effect on the soil was simply a consequence of their economic activity. But in some cases, the English took a more deliberate approach, becoming soil formers by design. Chapter 4 described how English farmers turned the sandy soils of barren heaths into productive farms during the 1700s and 1800s by mixing large amounts of clay into the surface. Chapter 4 also described how other English farmers made clayey land more tillable and more productive by mixing large amounts of chalk ($CaCO_3$) into the soil. This allowed water to move through the profile more easily and made cultivation much easier. Historical records show that this practice was followed in Britain for more than 2,000 years, continuing well into the 20th century.

San Joaquin and the Fresno Scraper

Intense modification of soil by humans is not limited to the British. Numerous examples can be cited from around the world, and one of the most interesting is the San Joaquin soil of California. Established in 1900, San Joaquin is the oldest continuously recognized soil series in the state, and it is now California's state soil. There are several reasons to discuss this soil. First, if a prize were awarded for the most interesting soil in the United States, San Joaquin would be a strong contender. Second, this soil is of great agricultural significance, having added billions of dollars to the California economy during modern times. Third, the things that humans do to San Joaquin soils in order to farm them are textbook examples of how soil can be transformed to meet human needs.

San Joaquin soils occur on nearly-level to undulating terraces on the eastern sides of the Sacramento and San Joaquin valleys. Under natural conditions, the soils have a distinctive micro-relief somewhat like minia-ture dune fields. Some have used the phrase "hog wallows" to describe them. The NRCS State Soils website has a very good landscape photo of this soil showing the rather strange looking surface configuration. Many of the shallow depressions between the hummocks fill up with water during winter and early spring, forming vernal or spring pools, which persist for a while and then dry out during the hot, dry summer that follows. And the summers are dry indeed. At the type location for the San Joaquin series near Lodi, California, only three inches or so of rain fall during the six month period from April through September.

Originally these soils were used mostly for livestock grazing or as unmanaged wildlife habitat, but they now are used to grow a wide variety of high value crops such as almonds, figs, pistachios, and grapes. But before these soils can be used for agriculture, they must be leveled to get rid of the uneven topography, i.e., the "hog wallows." Beginning in the early 1900s, a very useful invention known as the "Fresno scraper," pulled by mules or horses, was used to do this. In the late 1880s, a Scottish immigrant began manufacturing and selling "Fresno scrapers" in his local machine shop, and they soon became very popular.

But land leveling is only the first step. San Joaquin soils also have dense subsoil layers cemented by silica and/or clay that must be shattered in order to increase rooting depth and allow water to move down into the soil. Partial sticks of dynamite were used in the early years, but explosives soon were replaced by large tractors pulling slip plows. Slip plowing or ripping is accomplished by heavy tractors pulling long steel shanks through the soil. The shanks can shatter clay and silica-cemented pans to a depth of three to five feet, greatly increasing the rooting depth and making the soils suitable for orchards and vineyards. Fortunately, some areas of San Joaquin and similar soils, with their distinctive topography, hydrology and life forms, are being preserved in their natural state. Unlike earthworms, termites and other biological soil formers, humans can decide that in some cases we should restrain ourselves and allow certain areas to remain as natural and free of human impact as possible.

Topography and Soil Formation

Topography affects soil formation mostly by redistributing water and energy on the landscape. Assume a region receives 30 inches of rain yearly. Soils in nearly flat or undulating upland areas (e.g., broad ridges) will get the regional average of about 30 inches, the amount that falls from the sky, no more and no less. Since water runs off of sloping land, the adjacent side slopes will get 30 inches of water minus any that runs off down slope. In contrast, low-lying areas will get the regional average of 30 inches plus any that runs on from the uplands. Because of these basic hydraulic facts, soils in upland areas tend to be well drained, while many of those in low-lying areas tend to be saturated or even ponded with water for the entire year or for part of it; and it is a well-known fact that organic soils often form in low-lying wet or flooded areas.

Histosols, by definition, are soils that form from decomposing plant tissue, generally under waterlogged conditions in swamps and bogs. As such, they are a good example of how important topography can be in soil formation. If you look at a soil order map of the United States, you will note that large concentrations of these soils occur in Minnesota, Wisconsin, and Michigan. In the South, Histosols are most abundant in eastern North Carolina, southern Louisiana, and throughout Florida. About 70 percent of the Histosols in the continental United States are in the northern, formerly glaciated areas bordering the Great Lakes, while most of the remaining

30 percent are in the low-lying coastal areas of the southeast. Reminder: You can access a map showing the distribution of Histosols in the United States by keying a phrase such as "soil order map US Histosols" into a search engine.

During the Quaternary ice age, much of what is now Minnesota was covered numerous times by massive glaciers. Temperatures then became warmer and soon the glaciers retreated north, but the receding ice left the state with a much-altered topography. Prior to glaciation, what is now Minnesota is believed to have had a well-developed system of rivers and streams that effectively removed water from the land surface and carried it away to the sea. The ice obliterated that system and, upon retreating, it left behind a wet, soggy landscape covered with innumerable lakes and bogs. Ever since the ice melted, streams have been busily cutting into this immature landscape, and in time an efficient drainage network will be reestablished. But today much of the precipitation is captured and held temporarily in ponds and bogs, and since many streams simply meander from lake to lake or bog to bog, only a portion of the total precipitation is carried out of the area. A region such as this is an ideal setting for Histosols to form. Not surprisingly, Minnesota has many bogs with thick deposits of organic matter that soil scientists classify as organic soils or Histosols. It takes hundreds or even thousands of years for such soils to form. The process by which living organisms can slowly turn a pond or lake into organic soil is an interesting one.

Peat bogs occur throughout the world and many of them are created by mosses in the genus *Sphagnum*. Mosses were among the earliest plants adapting to life on land, but lacking true leaves or true roots, they are restricted to moist environments and cannot grow very large. Sphagnum moss, for example, lives in saturated, treeless bogs; but in addition to living in them, it actually creates them. A few sphagnum plants in a pond or lake can keep growing and expanding their biomass until the lake or pond is turned into a bog, with nearly all of the water held by the ever increasing mass of living and dead sphagnum. As the amount of sphagnum in a body of water increases, the environment is made increasingly inhospitable to competing life forms. Sphagnum bogs are so low in oxygen that tree roots cannot survive. Low oxygen also limits the growth of bacteria, so decomposition is minimal.

Over time, the growth of sphagnum makes the environment more acidic, slowing or stopping the growth of bacteria that might break down organic matter. Because it decomposes so slowly, dead sphagnum keeps piling up, and even when dead it can hold up to 20 times its weight in water, maintaining the saturated conditions that enable the living sphagnum at the surface to keep growing. It might take hundreds or even thousands of years, but in time sphagnum moss can convert a pond into an organic soil more than two meters thick. Peat bogs and other Histosols account for only three percent or so of the Earth's land surface, but they store more carbon than all of the forests on Earth combined (Kleier 2017).

Histosols of the Southeast

Like northern Minnesota, much of eastern North Carolina is wet and soggy, with meandering, slow-moving rivers and large areas of swamps and marshes. Streams are far apart and only a few meters or so below the land surface. In addition, the land is very flat and mostly underlain by impermeable sediments. Under these conditions, excess water must either run off as overland flow (a very slow process in such flat landscapes) or evaporate. According to Diemer and Bobyarchick (2005), the wetlands that characterize this area are "creations of topography." This topography has created a lot of wetlands and a lot of organic soils. For example, the Great Dismal Swamp of eastern North Carolina and Virginia once covered roughly 2,000 square miles (more than 5,000 square kilometers). The term "Dismal" requires some clarification. At one time, all large swamps in eastern North Carolina were commonly referred to as "dismals," probably because they were so gloomy, dreary, and dark, and there are a lot of swamps or "dismals" in eastern North Carolina. For example, Robeson County, North Carolina, alone has more than 50 named swamps. Although glaciers never made it as far south as North Carolina, the Histosols common to the eastern part of that state, like those of northern Minnesota, can be attributed largely to the melting of ice; but the story is a little more complicated.

About 300 million years ago, Africa collided with North America, creating the supercontinent known as Pangea. This monumental collision also created the ancient Appalachians, a mountain range that rivaled the modern Himalayas in size. The resulting supercontinent lasted for only a short time in geologic terms. About 225 million years ago, Pangea began to break apart and by 200 million years ago, the Atlantic Ocean had begun to form. As Africa and North America moved apart the ocean widened, and vast amounts of sediment were washed eastward out of the towering Appalachians to form the Atlantic Coastal Plain. During the millions of years that it took for the Coastal Plain to take form, sea levels rose and fell many times. Today the Coastal Plain of North Carolina is about 100 to 140 miles wide and consists of a series of terraces that descend to the ocean like stairsteps, with each terrace marking the elevation at which the ocean stood at some time in the past (Daniels et al. 1984). The three terraces that are lowest in elevation and nearest the ocean are referred to collectively as the Lower Coastal Plain.

At the peak of the last ice age, the Atlantic Ocean was several hundred feet lower than today and the Coastal Plain was much wider, with the shore located much farther to the east. With the end of the ice age, the glaciers began to melt and sea levels rose. As the Atlantic rose in elevation and reoccupied the Lower Coastal Plain, it created a drowned coast, with flooded river valleys; slowly moving, tidally influenced streams; and broad, slightly elevated uplands with water tables at or near the surface for much of the year. Slow natural drainage and high water tables created ideal conditions for thick organic soils to form. For example, the soil survey of Hyde County, North Carolina (Gagnon 2001) shows that Histosols cover nearly one third of the county, and Hyde County is not atypical. The Histosols of Hyde and other eastern North Carolina counties are typically one to three meters thick, but depths of four meters or more have been recorded.

Soil Formation on Side Slopes

In addition to creating wet soils in low-lying areas, the flow of water off of sloping land affects soil formation in the higher parts of the landscape, where a good portion of the rain runs off. Hazleton, the state soil of Pennsylvania, is a good example. It occurs in about half of the counties in the state, covering about 1.5 million acres or about 600,000 hectares. Large areas of this soil also occur in nearby states. Hazleton is classified as an Inceptisol, the name coming from the Latin *inceptum* or beginning, implying that there is minimal profile development. Inceptisols commonly form in recently deposited sediments or in areas where unstable slopes or cold temperatures limit soil development.

Hazleton is classified as loamy-skeletal, which means it is a mixture of soil and un-weathered rock fragments, usually sandstone; and while nearby soils on more stable landscapes have argillic horizons, these horizons are lacking in the Hazleton series. Soil development of the Hazleton series is limited mostly by topography, since the soils occur on steep, unstable hillsides in the highly dissected terrain of the Appalachian Highlands. The formation of soils such as Hazleton, because of their position on slopes, is greatly influenced by gravity, which is constantly trying to pull everything downhill.

Imagine a heavy rain falling on such a landscape. Some of the water enters the soil, but part of it runs off over the surface. On the way down, this moving water dislodges soil particles and carries them along; this, of course, is soil erosion, and the steeper the topography, the more erosion. Surface erosion limits soil formation in at least two ways. First, the continued loss of soil from the surface reduces overall soil depth. Second, any water that flows away down slope is not available to percolate into the regolith and deepen the soil by breaking down rock structure. Soils that form from layered rocks on steep slopes commonly contain a lot of nonweathered fragments, and soil scientists often describe them as being loamy-skeletal or sandy-skeletal.

In addition to surface erosion, sloping landscapes are subject to the destabilizing phenomenon commonly referred to as creep. Land surfaces with any slope at all are constantly, imperceptibly sliding or "creeping" downhill because of gravity. If the soil becomes saturated by heavy rains, the mass can move quickly, resulting in a slump or even a landslide. Inceptisols such as Hazleton are common on steep slopes because soil formation is constantly being slowed down or interrupted by surface erosion and by an unstable regolith that is always moving downslope — slowly most of the time, but very rapidly on occasion. Creep takes place so slowly that we are rarely aware of it, but it is constantly reshaping the land, creating much of the complex topography that we see around us.

Natchez, the state soil of Mississippi, is another Inceptisol in which soil development is curtailed by erosion and creep on sloping topography. This soil forms in deep loess deposits on highly dissected bluffs along the Mississippi River and its tributaries. Like the Hazleton soils farther north, profile development in Natchez soils is limited by surface erosion and by the fact that gravity keep interrupting soil formation by causing the entire soil mass to slide slowly but inexorably downhill. Readers are encouraged to check out both Hazleton and Natchez on the NRCS State Soils website.

Time and Soil Formation

The Histosols of northern Minnesota and eastern North Carolina are approximately the same age, both having formed after the glaciers began to melt and sea levels rose, which tells us that deep organic soils can form within a few thousand years. A typical rate of accumulation in woody peats, such as those in North Carolina, is about one inch (2.5 centimeters) in 25 to 50 years (Falini 1965), with faster accumulation in the case of peat moss or marsh grass. Taking a conservative approach, let us assume that it takes 50 years to form one inch of organic soil. Under these conditions, a Histosol two meters (80 inches) thick can form in about 4,000 years. Of course, a lot can happen in 40 centuries. There might be extended droughts during which organic matter accumulation ceases, and periodic fires might consume the upper layers of soil, destroying centuries of progress; but taking such uncertainties into account, it still is safe to say that Histosols a meter thick or more can form in only a few thousand years.

That is one way to think about time in relation to soil formation: how long it takes for a certain kind of soil to form at a given location. For Histosols it is many hundreds or a few thousand years, but what about some of the other orders? An Entisol, the simplest of the soil orders, can form in a very short time — a matter of minutes or less. According to Jenny (1941), "As soon as a rock is brought into a new environment and acted upon...it ceases to be parent material and becomes soil. We are forced to conclude that young riverbanks, fresh loess mantles, etc., are soils, unless they are being deposited under the very eyes of the observer."

There is little doubt that silt carried by a river can become soil (an Entisol) almost as soon as it is deposited on a floodplain or terrace. If left undisturbed, that Entisol can become an Inceptisol, with some evidence of soil formation, in only a few years. With the passage of a few hundred years more, the Inceptisol can develop an argillic horizon and become an Ultisol or Alfisol. Depending on parent material and climate, sequences of soil formation with time such as those below can be imagined. Spodosols and Alfisols can form in hundreds to thousands of years, but it can take more than a million years for some Oxisols to form.

Parent Material ⇒ Entisols ⇒ Inceptisols ⇒ Spodosols
Parent Material ⇒ Entisols ⇒ Inceptisols ⇒ Alfisols
Parent Material ⇒ Entisols ⇒ Inceptisols ⇒ Ultisols ⇒ Oxisols

Self-Similarity and Soil Mapping

In 1967, mathematician Benoit Mandelbrot published an article in which he asked, "How long is the coast of Britain?" He then argued that the answer depends on how closely you look. Viewed from a distance, almost any coastline looks smooth; but if you zoom in, it becomes more jagged, more complex, and longer. He also found that if you zoom in on a small section of a coastline, it is very similar in appearance to a much longer stretch viewed from a greater distance. His conclusion concerning Britain's west coast, as stated in a *New York Times* interview, was that "The length of the coastline, in a sense, is infinite." Mandelbrot argued that coastlines, like so many other things in nature, are self-similar or fractal. For example, a large stream has tributaries that are themselves composed of tributaries; these smaller tributaries are composed of yet smaller tributaries; and so on. So many things in nature are like infinite versions of the Russian matryoshka or nesting dolls. Examples of fractal patterns in the world around us are almost limitless; they include snowflakes, clouds, ferns, lightning, and trees. The timing and magnitude of earthquakes, the rhythm of a human heartbeat, the way DNA is folded into a cell — all are fractal. In Mandelbrot's words, things such as coastlines, rivers, snowflakes and DNA are "statistically self-similar, meaning that each portion can be considered a reduced-scale image of the whole."

Following Mandelbrot's example, let us ask the following question: How many kinds of soil are there in the world? By now the answer should not surprise you. It depends on how closely you zoom in. Assume that an astronaut is orbiting halfway from the Earth to the moon. Assume further that someone has cleared away the atmosphere and then delineated the world's major soil regions on the globe. These delineations can be seen from space, and each can be identified by one of the twelve orders recognized in Soil Taxonomy. Our astronaut would be justified in saying that, from his or her perspective, there appear to be twelve distinct kinds of soil in the world, each representing a unique combination of parent material, climate, organisms, relief, and time. Of course, there would be more than twelve delineations on the globe. For example, large areas of Mollisols occur both in North America and in Eurasia; while South America and Africa both have large regions dominated by Oxisols; and Aridisols or desert soils cover distinctive areas of Asia, North America, South America, Australia, and Africa.

But if our astronaut zooms in closer, he or she will observe that the factors of soil formation begin acting at finer and finer scales, carving the Earth's surface up into ever more intricate patterns of landscapes and soils. Perhaps the ultimate in zooming in is a detailed soil survey in which soil scientists walk across the landscape day after day, looking down at the surface from a height of five feet or so. In the United States, such maps are commonly prepared at a scale of 1:24,000, or one inch equals 2,000 feet. At this scale, contrasting soil areas as small as one hectare (about 2.5 acres) can be reliably identified. Such maps have now been completed for much of the United States and are widely used by farmers, tax assessors, environmentalists, conservationists, and others tasked with making land use decisions. Following is a case study of such a project.

A Case Study

In the early 1980s I served as project leader for the soil survey of Cumberland and Hoke Counties, North Carolina (Hudson 1984). We chose to map the two counties together largely because Fort Bragg, a sprawling military base, occupies large parts of both counties, and the United States Army was helping to pay for the project. Geologically the soil survey area is part of the Southeastern coastal plain and consists of ancient, ocean-deposited sediments 100 to more than 400 feet thick over volcanic rock. Of course, the volcanic rock is buried so deeply that it plays no role in soil formation. Through field investigations, we determined that the survey area, which was 666,880 acres in size, could be mapped using 48 soil series; in other words, there were 48 distinct, mappable soil units in the survey area that resulted from 48 unique combinations of climate, organisms, relief, parent material and time (cl, o, r, p, t).

For more than a century, soil series have been the category of choice for making soil surveys in the United States, and many of them have a long history. The table below shows how long some of our most venerable soil series have been around. Two series in the list, Cecil and Windsor, were established prior to 1900. Soil series are the lowest category of Soil Taxonomy, in which the hierarchal arrangement is Order > Suborder > Great Group > Subgroup > Family > Series. Series, of course, came first; the higher categories were introduced later, an implicit recognition that the factors of soil formation are fractal, varying with geographic scale.

Table 8.2. Selected state soils and dates they were established

State	State soil	Date soil series established
California	San Joaquin	1900
Connecticut	Windsor	1899
Indiana	Miami	1910
Maryland	Sassafras	1900
North Carolina	Cecil	1899
Texas	Houston Black	1902

Soils are grouped together in the same series if their properties are enough alike that they will perform similarly for most land use purposes. More specifically, a soil series has horizons that are similar in composition, color, texture, structure, reaction (pH), consistence, and other properties that affect use and management. The soils in a series also must be similar when it comes to natural drainage. For example, a soil in which the water table remains near or above the surface during the winter cannot be in the same series as one in which the water table remains below a depth of two meters throughout the year. Because soil series by design have a narrow range of properties, all of the soils in a series are similar in terms of behavior and potential for use. That is one of two reasons why soil series are so useful in detailed soil mapping. We will get to the second reason shortly, but first let us return to the Cumberland–Hoke soil survey, with its 48 soil series.

It might seem that 48 soil series are a lot to remember and to map out on the landscape, but the task was quite manageable for several reasons. First of all, the series were not distributed randomly across the survey area; instead, they tended to cluster in certain locations. For example, one large section of the survey area, called the Sandhills, was found to consist of broad, sandy ridges dominated by Entisols and less sandy side slopes dominated by Ultisols. Incidentally, the broad sandy ridges are the main reason that Fort Bragg is located there, since they provide a soft landing for young soldiers jumping out of airplanes. Streams have cut deeply into the sediments, so the sandy uplands drain rapidly, even during extended wet periods. As a result, nearly all of the soils are well to excessively drained. We were able to map nearly all of the Sandhills area using just five soil series, and it did not take long for mappers working there to learn how to recognize them.

We found that a small number of soils also clustered on the broad meander belt of the Cape Fear River, which flows north to south through the survey area. This belt, five to eight miles wide, consists of a system of terraces and floodplains laid down by the river as it has meandered across the landscape over a period of many centuries. This section of the survey area was mapped using only seven soil series. Again, it did not take long for members of the soil survey crew to learn what each of these soils looked like and to understand how to tell them apart. On a given day or during a given week, a mapper on the meander belt could expect to encounter seven soils at most, but often fewer. Distinctive soils also clustered in other sections of the survey area, and because of this geographic partitioning, new project members were able to familiarize themselves with soils a few at a time, greatly simplifying the learning process.

But learning to recognize the soils and tell them apart was just the first step. Soil mappers then had to master a more difficult skill, learning how to draw maps showing how the different soils are distributed on the landscape. It is feasible to do this because, in stable land areas that have evolved under a protective vegetative cover (as was the case in this soil survey area) there is a strong correlation between soil series and natural, mappable parts of the landscape. The predictive power of this relationship enables soil scientists to delineate bodies of soil accurately.

To understand how this works, one must put aside an idea that is widely espoused, the idea that soil is a continuum on the landscape. Soil does behave as a continuum over short distances. However, the soil cover is interrupted by frequent, often abrupt discontinuities that can be seen and mapped out on the landscape by trained observers. The term discontinuity, as used here, refers to a boundary area on the landscape where one or more of the soil-forming factors changes rapidly within a short distance. A change in one or more soil properties typically occurs at the same location and within the same lateral distance. These abrupt soil changes at observable discontinuities make soil mapping possible.

This brings us to the concepts of landform. Landforms are the observable features that make up the Earth's land surface. They have a characteristic shape and include large features such as plains, plateaus and mountains. They also include smaller features such as dunes, drumlins, ridges, side slopes, and stream terraces and floodplains along streams of any size. Landforms are recognized by their shape, by their position in relation to other landforms, and by the kind of geologic material in which they form. Landforms, like most natural features, are fractal or self-similar; how many you see depends on how closely you look, and this is where the discussion gets a little complicated. Landforms are not uniform; for example, an upland might consist of convex areas or swells in the higher parts of the landform alternating with flats or depressions on lower parts of the landform. Mandelbrot and his disciples would argue that these simply are smaller landforms that we can observe because we are looking more closely. But many geologists would refer to these smaller features as landform components. When mapping soils, we usually are interested in mapping out these landforms or, in some cases, landform components, because we have learned that each landform or component typically has its own distinctive soil. For example, in nearly all cases, the soil on a very steep slope is different from the soil on a gentler slope, so the very steep slope would be a different landform component than the gentler slope. Similarly, convex swells and low-lying swales on uplands typically have different kinds of soil. A basic knowledge of landforms and landform components enables one to understand the soil mapping process (Hudson 1992).

- Within each landform or component, the five factors of soil formation interact in a distinctive manner. Each landform or component results from a unique combination of climate, organisms, relief, parent material, and time. As a result, all areas of a given landform or component develop the same kind of soil, which typically is identified by a soil series.

- In a given soil survey area, the total number of landforms or components and thus the total number of soil series is limited, and each distinct landform component-soil series combination appears again and again as one traverses the landscape.

- The more dissimilar two landform components are, the more dissimilar their associated soils tend to be. Conversely, landform components that are similar tend to have soils with similar properties.

- The more dissimilar two adjoining landform components are, the more abrupt the discontinuity separating them. An example is a steep hillslope and a nearly level floodplain or terrace at its base. Conversely, the more similar two adjoining landform components are, the less abrupt the discontinuity separating them. This means that adjoining areas with the most contrasting soils usually can be separated most accurately and precisely on the landscape.

- Adjacent areas of different landform components usually have a predictable spatial relationship one to another. For example, one kind of component will always be above another on the landscape, or between another and a stream.

- Once the correlations between soils and landform components have been determined and the spatial relationships among landform components become known for a soil survey area, the soils can be mapped by identifying and delineating their associated landform components. The soil is examined directly only as needed to validate the underlying relationships.

The near equivalency of soil series and landforms (or, in some cases, their components) makes soil mapping feasible; the fact that soil series and landforms are so highly correlated has made it possible for the soil factor equation to become the driving force behind a large, enduring technical program. Using this equation, which represents a simple but powerful concept, the National Cooperative Soil Survey of the United States has now made detailed soil maps and accompanying data available for more than 90 percent of the nation's land area, with the goal of reaching 100 percent in the very near future. But why did they do this? Why do we make soil maps and why do people want them?

To address these questions, let us return briefly to the soil survey of Cumberland and Hoke Counties, North Carolina. Although the survey area was and is undergoing rapid urbanization, agriculture and commercial forestry will remain important economic activities for the foreseeable future; and when it comes to these enterprises, soil productivity matters. For example, we found that the four most productive soils in the survey area could produce, on average, nearly twice as much corn or soybeans each year as the four least productive soils. For corn, the yield was more than 120 bushels per acre on the best soils versus about 65 bushels per acre on the poorest soils; for soybeans, the relative yields were 45 and 25 bushels per acre respectively. This did not include six soils in the survey area that were rated as totally unsuited for agriculture. If you are a farmer, a land appraiser, someone buying or selling farmland, or even a local official wanting to tax land fairly, this is the kind of information you need. Note: Crop yields given here reflect the data circa 1980. Yields could be higher now due to improvements in agricultural science and farming technology.

Commercial forestry, which in much of the southeastern United States involves the cultivation of loblolly pine (*Pinus taeda*) for pulpwood and timber, also is important to the local economy. Foresters typically measure the productivity of forest land in terms of site index, which is the height a tree is expected to reach in 25 (Site Index 25) or 50 years (Site Index 50). American foresters usually express site index in feet, but they occasionally use meters. On the four best timber growing soils in the survey area, loblolly pines grew, on average, to a height of 97 feet (30 meters) in 50 years. In contrast, loblolly pines on the poorest sites reached a height of only about 60 feet (18 meters) in 50 years. On average, trees grew 1.5 times as fast on the best soils as on the poorest. Although foresters measure tree heights, they really are interested in tree volume, how much usable timber or wood fiber they can produce on a given piece of land. If we think in terms of volume instead of height, the differences between the best and poorest soils become even more pronounced. In our case, the average 50-yr-old loblolly pine on the best sites contains nearly 90 cubic feet of wood. In contrast, the average 50-yr-old loblolly pine on the poorest sites contains less than 25 cubic feet. On a volume basis, the best soils are capable of producing about 3.5 times more wood than the poorest soils. If you are in the business of managing, buying, selling, appraising, or taxing forest land, this is the kind of information you need to know.

There is little doubt that soil maps are of value on land used for agriculture and forestry. But how useful are they in urban and residential settings? In the Cumberland–Hoke soil survey, we found that 31 of the 48 soils mapped, accounting for more than one third of the survey area, were unsuitable for building houses or for most urban uses. Some of the soils had a high water table, some were subject to flooding, some ponded after heavy rains, and some were steep and unstable. Anyone building on such soils would likely suffer financial losses, in some cases severe. On the other hand, 13 of the 48 soils mapped, accounting for about 46 percent of the survey area, were well suited to residential and urban development. Anyone building houses or commercial buildings on these soils could expect to encounter few problems. If you are a homebuilder, a county or city planner, or a banker looking to lend money for residential or urban development projects, this is the kind of information you need.

Costs and Returns of Soil Surveys

Klingebiel (1966) compared the costs and returns of soil surveys in different settings. He calculated the benefit to cost ratios of soil surveys in range and forest land, in areas of mixed agriculture (about half cropland), and in areas near rapidly growing towns and cities. Expressed another way, he compared the costs and returns of soil surveys under conditions of low, medium, and high use intensity. The numbers he came up with are rather impressive; he concluded that the average benefit to cost ratio, or return on investment, for soil surveys was about 40:1 on range or forest land, about 60:1 on agricultural land, and well over 100:1 in rapidly expanding urban areas. He concluded that the entire cost of most soil surveys is recovered very quickly, usually within a year or two; in his words, "This indicates a minimum of nearly $2 of return the first year for each dollar spent for the entire cost of the soil survey."

Klingebiel cited some specific examples, including a town in Massachusetts where officials saved $250,000 (in 1966 dollars) by using soil survey information to plan a new sewage system. The state of Illinois, by using soils information, was able to save $810,000 dollars in construction costs on just one ten-mile stretch of interstate highway. A farmer in Fayette County, Tennessee, used soil survey information to increase his income by $5,550 in one year alone, while farmers in Hall County, Nebraska, by using soil surveys to improve water management, increased their annual incomes by $5 to $150 an acre. Klingebiel also cited examples of expensive mistakes resulting from not using soils information. Officials in one county incurred

$250,000 in avoidable costs when they selected a site for a new school without consulting a soil survey. A location just 500 feet away would have been much better. Klingebiel concluded, "A soil survey is an investment that is almost certain to pay for itself — and return a profit — within a year." Soil surveys, as a rule, provide very high returns on investment. Accurate, very cost effective soil maps are possible because of the fractal or self-similar nature of soil variation and the fact that mappable soil areas and natural landforms are so highly correlated.

CHAPTER 9. HOW SOILS EVOLVED

Soils have been forming on Earth for a very long time. Sedimentary rocks 3.8 billion years old have been found in Greenland, and the presence of such rocks is evidence that rain, running water, lakes and oceans existed at the time (Allaart 1976). Scientists are reasonably certain that the atmosphere at that time contained carbon dioxide, so hydrolysis, the main process by which rocks are broken down, was taking place then much as it does today (Retallack 1990). All sedimentary rocks start out, of course, as sediments, and nearly all sediments form soils at their surfaces soon after being deposited. This is pretty strong evidence that both the climatic conditions and the parent materials present prior to 3.5 billion years ago were conducive to the formation of soils, albeit primitive ones by modern standards. There is more direct evidence in the form of paleosols or ancient buried soils. Scientists have found one paleosol that is at least 3.4 billion years old (Lowe 1983) and the oldest paleosol studied in detail was determined to be about three billion years old (Grandstaff et al. 1986).

We can infer that any soils older than 3.5 billion years were either Entisols, Inceptisols, or Andisols because environmental conditions did not yet exist for the other soil orders to form. For example, Gelisols are perennially frozen soils that form under very cold or periglacial conditions. Since the first evidence of glaciation was around 2.3 billion years ago, Gelisols must date from about that time; they could not have begun forming earlier. The same is true of Histosols, which form from dead plant material. Since no land vegetation existed prior to about 470 million years ago, the first Histosols must date from that time; they could not have evolved earlier. The remaining soil orders can be eliminated in a similar manner.

Several lines of evidence suggest that Entisols, Inceptisols, and Andisols evolved prior to 3.5 billion years ago. All it took to form primitive Entisols and Inceptisols were rocks, weathering agents to break them down, and running water to carry away small rock particles and then redeposit them. Water-borne particles of sand, silt, and clay would have become soil (Entisols) soon after they were deposited on the banks of a river or stream, and some of these Entisols would have become Inceptisols in a very short time. But Andisols are different; they require a specific kind of parent material — volcanic ash or similar materials ejected from volcanoes. We know that if such materials are exposed at the surface in a humid environment, they weather rapidly to form glass and colloidal materials that have what Soil Taxonomy describes as andic properties. Andisols are generally considered young soils because ash or similar volcanic ejecta weather so rapidly when exposed to air and water.

The climatic conditions under which Andisols form were present well before three billion years ago; there was rain, the atmosphere contained abundant CO_2, and there is evidence that the right kinds of parent materials existed also. No large continents existed at that time, but there were many active volcanos and volcanic islands. According to Retallack (1990), "The world in those days can be imagined as an Indonesian archipelago of global proportions... Massive spines and walls of volcanic intrusions may have towered over undulating clayey terrain of deeply weathered ash and lava." This sounds like an environment in which Andisols could have formed.

In addition to Entisols, Inceptisols, and Andisols, another soil was present at that time — one that no longer exists. The mysterious "Green Clay," as pedologists refer to it, appeared sometime before three billion years ago and persisted for more than a billion years before going extinct. At the time these "Green Clays" formed, there was no free oxygen in the atmosphere. When photosynthetic organisms started pumping oxygen into the air millions of years later, these early soils were doomed, since they could not persist in an oxygenated atmosphere. As far as we know, this is the only major soil group to evolve on Earth, persist for a long period of time, and then go extinct.

Paleopedology

Geologists have been aware of paleosols or ancient buried soils for more than 200 years. In 1795, James Hutton, one of the founders of geology, referred to the *sol mort rouge* or "dead red soil" often seen at the boundaries between coal seams in mining areas of Germany. During the 1800s, distinctive paleosols were identified in Russia, the American Midwest, and in New Zealand. Russian scientist B.P. Polynov introduced the term "paleopedology" to the scientific literature in 1927 and another Russian, C.C. Nikiforoff, introduced the concept of paleosols to American soil scientists a few decades later. Nikiforoff, after completing his scientific studies in Russia, immigrated to the United States and began working for the Soil Conservation Service. The ability to recognize and date ancient soils has proven useful in geology, climatology, archeology, and other fields. As a result, interest in the discipline has grown, techniques have improved, and the knowledge base continues to expand. For a comprehensive, rather technical discussion of paleosols, see G.J. Retallack (1990), *Soils of the Past: An Introduction to Paleopedology*.

Retallack points out that in many cases, buried, fossilized soils can be placed in one of the orders of Soil Taxonomy. He asserts that, although Soil Taxonomy was not designed with paleosols in mind, its categories "are serviceable for paleopedological studies." Paleosols can most often be identified by the very sharp boundary between the overlying material and the surface of the buried soil. Other indicators include visible soil horizons, fossilized roots, old root channels, and animal burrows. Paleopedology is still a young science, and it is difficult to find and identify soils that have been buried for millions and sometimes billions of years. However, a number of very old paleosols have been classified and dated, providing us with at least some supporting evidence for the following timeline, which shows when each of the twelve orders in Soil Taxonomy evolved.

Table 9.1. Timeline for the evolution of soil orders

Soil Order	Time of appearance on Earth
Entisols, Inceptisols, Andisols	3.8 to 3.5 billion years ago
Gelisols, Vertisols, Oxisols, Aridisols	2.5 to 1.5 billion years ago
Histosols, Spodosols, Alfisols, Ultisols	470 to 360 million years ago
Mollisols	65 to 50 million years ago

The timeline above is very conservative; some of the orders probably evolved earlier than shown in the table. However, we can be reasonably certain that each order had become well established by the end of the time interval indicated above. It is clear that the soil orders did not appear one at a time or all at once; instead, for the most part they arrived in groups, with long intervals of time between the groups. This is a classic example of punctuated equilibrium, a concept introduced by Eldridge and Gould in 1972. For a long time, it was thought that evolution came about entirely as a result of slow, incremental changes taking place over very long periods of time, but Eldridge and Gould proposed a different model. In their opinion, the evolutionary process can be characterized by long periods of stasis or equilibrium punctuated at intervals by sudden onsets of rapid environ-mental change during which new species arise and some old species die out. After each flurry of activity, a new stasis or equilibrium is established. The new equilibrium might last a very long time, but eventually something will disrupt the balance once again, setting the stage for yet another burst of evolutionary activity.

The rest of this chapter presents evidence to validate the timeline above, while at the same time adding detail and context. In the process, it will become clear that soil evolution is almost a textbook example of punctuated equilibrium. As mentioned earlier, the first three soil orders — Entisols, Inceptisols, and Andisols — probably evolved prior to 3.5 billion years ago and almost certainly were in place before 3.0 billion years ago. The now extinct "Green Clay" also evolved at that time. This was followed by a period of equilibrium or stasis lasting more than half a billion years. Then, within a relatively short time, four new soil orders appeared.

By 2.5 billion years ago, small continents had begun to form and some parts of them soon became glaciated. According to Retallack (1990), "The amalgamation of continents of some size and elevation is suggested by peri-glacial deposits as old as 2,300 million [2.3 billion] years. Long-buried ice wedges and other features indicate that the order Gelisols dates from this time period (Young and Long 1976). In addition to Gelisols, the first Verti-sols began forming in the interiors of small continents sometime before 2.0 billion years ago. Geologists examined a 2.2 billion-year-old paleosol and found that it had properties very similar to those of modern Vertisols (Retallack 1986). Therefore, there is good evidence that both Gelisols and Vertisols evolved sometime between 2.5 and 2.0 billion years ago.

Oxisols and Aridisols also appeared at about this time. As continents grew larger, deserts began to form, especially in areas of rain shadow created by high mountain ranges. Retallack (1990) puts a fairly precise time to the appearance of Aridisols. "Extensive desert dunes, calcareous soils, and silcretes did not appear until ca. 1,800 million [1.8 billion] years ago." In addition to continent formation and glaciation, another process was occurring that would set the stage for the appearance of yet another soil order. Rye and Holland (1998), by measuring the iron content of fossil soils, determined that a sharp rise in atmospheric oxygen took place between 2.2 and 2.0 billion years ago. Although soils showed some evidence of oxidation prior to 2.0 billion years ago, the oldest soil clearly identified as an Oxisol is about 1.7 billion years old (Chiarenzelli et al. 1983; Stanworth and Badham 1984). There is good evidence that Gelisols, Vertisols, Aridisols, and Oxisols all evolved sometime between 2.5 and 1.5 billion years ago. Gelisols and Vertisols probably appeared shortly before 2.0 billion years ago; Oxisols and Aridisols probably appeared shortly after 2.0 billion years ago.

The Soil Factor Equation

We now have accounted for seven of the twelve soil orders and for approximately two billion years of soil evolution. Before proceeding, it will help to revisit the soil factor equation, a topic covered at some length in Chapters 7 and 8. This equation identifies the five formative factors that determine the properties of any soil anywhere on Earth. The soil factor equation was conceived independently by Dokuchaev in Russia during the late 1800s and a few years later by Hilgard in America. The soil factor equation was validated by Jenny and others and became the guiding force behind a very successful soil survey program carried out by the National Cooperative Soil Survey of the United States during the 20th century. According to this functional equation, the nature of the soil at any location on Earth is a function of climate, organisms, topography, parent material and time, or in symbolic form:

$s = f(cl, o, r, p, t)$.

Chapters 7 and 8 demonstrated the power of this equation in explaining how soils vary across the landscape, i.e., how they vary in space. This equation is equally useful in understanding how soils have varied over time. Prior to 3.0 billion years ago, only Entisols, Inceptisols, and Andisols existed because the available combinations of climate, relief, and parent material at that time could lead only to those three soils. But during the next billion years, oxidation, glaciation, and the amalgamation of small land areas into larger continents ushered in new kinds of climate, new kinds of topography, and new kinds of parent material, bringing forth new combinations of cl, r, p, and t and making it possible for four new soil orders to evolve. As of 1.5 billion years ago, seven of the twelve soil orders — Entisols, Inceptisols, Andisols, Gelisols, Vertisols, Oxisols, and Aridisols, — were in place.

More than a billion years went by before the equilibrium once again was disrupted and four additional soil orders appeared. The new soils appeared as a result of vascular plants evolving on land, covering the Earth's surface in greenery, and developing deep roots. This colonization began about 470 million years ago, but plants did not begin to have a pronounced effect on soil formation until about 400 million years ago with the evolution of lignin. Lignin added rigidity to stems, making it possible for plants to grow very tall and to produce extensive, deeply penetrating root systems.

The Age of Modern Soils

Mosses, the first terrestrial plants, began colonizing land surfaces as early as 470 million years ago. Lacking roots, stems, or other vascular tissue, mosses grew in moist, low-lying areas or in locations receiving large amounts of rain. Since nearly every cell in a moss plant can carry out photosynthesis, there was no need for a vascular system; individual cells were able to stay hydrated and obtain mineral nutrients by absorbing water that partially covered the plants or flowed over them during wet periods. Mosses soon were joined by fungi, and these two lifeforms shared the world's land surface for millions of years. Lacking roots, mosses had minimal effects on rock weathering or soil formation; nevertheless, as moss plants died and decayed, they added organic matter to the soil surface. For this reason, the shift from four factors of soil formation (cl, r, p, t) to five factors (cl, o, r, p, t) is dated at 470 million years ago, the time when mosses first colonized the land (Lenton et al. 2012).

As generation after generation of mosses died, their remains were preserved in wet low-lying areas where poor drainage and lack of oxygen slowed the decomposition of their remains. As a consequence, thick layers of partly decomposed organic matter began accumulating on the ground surface, soon resulting in the appearance of a new soil order, Histosols. We cannot be sure when the first Histosols appeared, but conditions necessary for their formation probably were in place shortly after 470 million years ago.

Ferns were the next major plant group to evolve on land, possibly as early as 430 million years ago (Testo and Sundue 2016). Unlike mosses, ferns have vascular tissue to transport water and nutrients from the soil up into the plant. Roots, leaves and stems enabled ferns to increase the amount of sunlight captured and the amount of water taken up from the soil. Ferns thrived, spread across the land, and soon began to grow very tall. They were able to do this because of lignin, a tough, carbon-rich polymer that evolved around 400 million years ago. The great strength of lignin enabled ferns to produce woody trunks, lift their foliage high above the ground, and in time form the world's first forests.

Geologists call the time from 419 to 359 million years ago the Devonian Period. The Devonian also could be called the Age of Modern Soils. The first small ferns are believed to have evolved on land about 430 million years ago, followed by the evolution of lignin about 30 million years later, enabling ferns to form woody trunks and grow tall and straight. The first true trees, with lignin and conifer-like wood, appeared about 15 million years later, and by 360 million years ago, magnificent forests had covered much of the world's land surfaces. We can, somewhat arbitrarily, set the beginning of the Age of Forests and the Age of Modern Soils at around 400 million years ago, roughly coinciding with the evolution of lignin. Lignin made it possible for terrestrial plants to produce large amounts of biomass and for individual organisms (trees) to grow massive and tall, requiring deep, extensive root systems. Below is the timeline for the evolution of the first forests, which closely parallels the evolution of modern soils.

470 million years ago: Mosses began colonizing land surfaces.
430 million years ago: Ferns, the first vascular plants, evolved on land.
400 million years ago: Lignin evolved.
385 million years ago: The first true trees with conifer-like wood appeared.
360 million years ago: Forests covered most land surfaces.

In nature, a change in one part of an ecosystem can sometimes affect other parts that at first glance seem far removed or unconnected. The evolution of vascular plants with deep roots is an example of this. To imagine what most landscapes looked like before the advent of terrestrial vegetation, think of how deserts look today. The land typically is eroded and gullied and the streams, although choked with debris, remain dry much of the year. Now think of how some of these same unprotected desert landscapes would look if they received 40 inches (a meter) or even 60 inches (1.5 meters) of rain annually. Before the land was covered with vegetation, nearly all of the rain ran off everywhere, causing catastrophic erosion and gullying. Streams had numerous channels that split off and then rejoined, giving them a braided appearance. Streams assume this form when there is more sand and gravel in the runoff water than can be carried, and the excess is deposited as islands between channels. The stream then spreads out, flowing on top of previously deposited material. Braided streams still are common in deserts and in areas receiving meltwater from glaciers.

Early forests reduced such erosion dramatically, landscapes became more stable, and rocks began to weather more deeply. Instead of running off of the surface as before, most of the rain percolated down into the soil. Much of it was quickly taken up again by tree roots and returned to the atmosphere, but some reached the groundwater and then moved slowly to nearby streams. Because of the slow, more controlled release of water, stream morphology changed. Braided streams that flooded after each rainstorm and then were dry the rest of the year gradually gave way to meandering rivers that flowed year round. The shape and flow rate of streams and rivers changed dramatically, but the most significant changes took place on the land itself, and nearly all of these changes were due to the transformative power of roots.

At first glance roots appear simple, but close study reveals that they are surprisingly complex in both structure and function. Roots first evolved around 400 million years ago and by 375 million years ago had penetrated a meter or more into the substratum. Below is a definition of roots from a scientific paper (Raven and Edwards 2001).

> Roots are axial multicellular structures of the sporophytes of vascular plants which usually occur underground, have strictly (root cap notwithstanding) apical elongation growth and generally have gravitropic responses which range from positive gravitropism to diagravitropism, combined with negative phototropism.

This undoubtedly is a good scientific definition, but it is highly technical and assumes a lot of previously acquired or a posteriori knowledge, so much so that those of us who are not botanists can hardly comprehend its meaning; but that is not really a problem. Anyone who can read and understand this book already has a basic understanding of what plant roots are and what they do. For our purposes, we need only add the context and scientific detail needed to fully appreciate how important plant roots have been to the evolution of modern soils.

If you could somehow measure the length of all the roots of a mature tree, the total would likely exceed several hundred miles, but only a small section at the very end of each root carries out the important tasks of absorbing water and nutrients. A tiny, hard cap at each root tip functions much like a drill bit as it bores through the soil. If the tip encounters a rock or other hard object, it simply makes a detour and continues on its way. The penetrating roots, or rootlets, become very small near the tips, approaching the diameter of a fine piece of twine. Just behind the root tip, thousands of root hairs grow out at right angles; their job is to absorb water and nutrients that are then transported up into the plant. Each root hair is formed from a single specialized cell and is typically less than one fifth the diameter of a human hair. Root hairs live only a few weeks before dying off, with replacements quickly emerging just behind the tip as it advances through the soil. In addition to absorbing water and minerals, root hairs anchor the advancing root in place, helping it to penetrate the soil more efficiently; as a result, root hairs usually are coated with soil particles. Root tips also have a gob of mucilage or a "slime film" which lubricates the tip. The existence of these films was predicted as early as 1883, but scientists were unable to see them at the time; but new staining agents and higher magnification soon made it possible to produce stunning pictures of them. How fast do roots grow? Weaver (1926), based on extensive studies of prairie grasses in Nebraska, concluded that under favorable conditions, "over half an inch a day" is not uncommon.

In addition to producing its own lubricant, growing root tips release a swarm of H^+ ions to help break down the surrounding rock; in the words of Keller (1957), roots "carry negative charges on their surfaces and are surrounded in the soil by an atmosphere which is composed mostly of H^+ ions." Scientists have measured the acidity on the root surfaces of some common plants such as cotton, soybeans and tomatoes and recorded pH values that were typically below 4.0, with some as low as 2.0. Proton pumps in root cells produce H^+ ions much like those in the walls of our stomachs produce the acid needed to digest food. The process is roughly the same, except that plant roots are digesting rocks instead of food. While roots are producing H^+ ions underground, decomposing litter at the surface is releasing H_2CO_3 and organic acids into the upper part of the soil, further increasing the intensity of rock weathering.

Forests that evolved during the Devonian greatly increased both the depth of weathering and its intensity. According to Berner (1997), "the dissolution of rock beds by weathering at this time was accelerated by the growth of plants...[A] sevenfold acceleration has been determined." Much of this acceleration can be directly attributed to roots, but not all of it. In addition to stabilizing land surfaces and transforming the nature of streams and rivers, vascular plants also brought about major changes in the hydrologic cycle, and these changes also accelerated rock decomposition. Chapter 3 pointed out that chemical rock weathering does not occur in the absence of water and emphasized that a single saturation is rarely enough. Field geologists see few signs of weathering in deeply buried rock that is continuously saturated with water. In contrast, rocks nearer the surface, where the water table fluctuates up and down during the year, typically show much stronger evidence of chemical weathering. Chemical rock weathering is favored by repeated infusions of fresh water to leach away the products of weathering as it passes through the rock

Evapotranspiration

Chapter 7 discussed the balance between evapotranspiration and precipitation in different regions of the United States and how important this balance is to soil formation. As the name implies, evapotranspiration refers to soil water that is lost to the atmosphere by simple evaporation plus that lost through transpiration. Transpiration is the process by which plant roots absorb water from the soil and transport it up to the leaves, where it then is lost to the atmosphere by evaporation from tiny pores. Plants transpire a lot of water; according to Black (1957), "Several hundreds of pounds of water may be lost by transpiration during the production of a single pound of dry matter." Plants, in order to meet the needs of photosynthesis, must transpire a lot of water just to keep their tissues hydrated, and this has turned out to be a very good thing. the prodigious amounts of water transferred to the atmosphere by terrestrial vegetation during photosynthesis are now an important component of the modern hydrologic cycle. The last sentence specified modern hydrologic cycle, because transpiration did not become an important part of nature until forests began to evolve shortly before 400 million years ago. Evaporation was taking place at the time, but very little transpiration.

This raises an important question: How much of the water loss from land surfaces is due to evaporation and how much of it is due to transpiration by plants? Frank J. Veihmeyer, who worked at the University of California at Davis for many years, conducted some simple but revealing studies on this topic during the early part of the 20th century. In one experiment, he added a known quantity of water to a four-foot (1.2-meter) column of soil in a tank with a closed bottom. The soil was kept bare of any vegetation and placed out of doors for four years. During the four years, no rain was allowed to fall on the surface. The tank was weighed at intervals to determine how much water had evaporated. At the end of the four-year period, only one fourth of the original water had been lost through evaporation, and nearly half of that loss had occurred during the first three months (Veihmeyer 1927). This simple experiment showed that in the beginning, evaporation from a soil surface is nearly identical to that from a free water surface. But after a thin, dry layer has formed at the surface, the rate of water loss from the entire soil is greatly diminished. This is the principle behind the alternate crop, fallow systems of agriculture commonly used in semiarid regions. Penman (1941) and others later replicated Veihmeyer's results using more elaborate experiments.

In addition to assessing the rate of evaporation from a bare soil, Veih-meyer conducted another interesting experiment. This time he added known amounts of water to two four-foot columns of soil in tanks with closed bottoms; he left the soil in one tank bare, while seeding the other with vetch (*Vicia* spp.). Both tanks were then placed out of doors. They were covered during rains and no additional water was added. When the vetch had grown to maturity, the amount of water lost from each tank was determined. Veihmeyer found that the soil with vetch growing in it had lost more than eight times as much water as the bare soil. This answers the question posed above; most of the water lost by evapotranspiration is due to transpiration by plants; only a small percentage is evaporated directly from the soil.

Before vegetation covered the Earth, most of the water falling on land surfaces ran off, and the small amount that made its way into the soil remained there for a long time. A very thin surface layer would dry out quickly, but then water loss from the soil below would be very slow. The evolution of deep-rooted forest trees changed all of that. The Devonian forests took root a little after 400 million years ago and after that there was much less runoff and much less erosion, because most of the rain that fell to the ground began percolating into the soil. Much of it was quickly taken up by tree roots and returned to the atmosphere by way of evapora-tion from leaf surfaces (transpiration). Thus began the modern hydrologic cycle on land:

Rain falls to the ground.

The rainwater percolates into the soil where much of it is taken up by plant roots.

Water is transported up to the leaves, where it evaporates into the atmosphere.

With time and cooling, water vapor condenses and falls to the Earth again as rain.

The rainwater percolates into the soil, beginning the cycle again.

The modern hydrologic cycle began soon after the first forests evolved and continues to this day. The initiation of this cycle coincided closely with the appearance of modern soils, and that did not happen by chance. In addition to stabilizing the land, Devonian forests began depositing large amounts of organic matter in and on the soil. This resulted in softer, more permeable soil surfaces that allowed for faster infiltration of water. At the same time, trees began transpiring large amounts of water, increasing the quantity of water cycled from the soil to the atmosphere and back as well as the frequency with which it cycled. As more water made more trips to the atmosphere and back, the total contact time between rocks and water became greater, accelerating the rate of chemical rock weathering and clay formation.

Most of the world's soils have mineral A horizons with varying amounts of organic material or humus mixed in. The mixing of decomposed organic matter into the soil surface, which began around 470 million years ago and became more pronounced after 400 million years ago, was a pivotal event in Earth's history. It was a necessary prelude to the establishment of the modern carbon cycle. Chapter 6 pointed out that the atmospheric supply of CO_2 is very small; out of every 10,000 atoms or molecules in the atmosphere, only about four are CO_2. In order to keep life going, this small supply of carbon dioxide must be recycled and reused again and again. The A horizons of soil play an important role in that recycling process. World-wide, soils contain an estimated 1,500 gigatons of carbon, about twice the amount stored in the atmosphere or in living plants. Each year, the amount of new carbon coming into the soil in the form of dead plant and animal material is roughly balanced by that returned to the atmosphere as older soil organic matter is oxidized.

While creating organic soils and helping to start the modern carbon cycle, terrestrial vegetation also altered the nature of mineral soils. The distinct A horizons that began forming at the surface of mineral soils in the Devonian came about by way of a soil forming process commonly referred to as humification. Humification takes place when organic matter deposited in or on the soil surface is decomposed by microorganisms to form humus. Humus is dark brown or black, giving A horizons their distinctive colors. In areas with good drainage, organic matter decomposes and forms humus rapidly; insects and other invertebrates carry organic matter below the surface, creating an A horizon in which organic matter in various stages of decomposition becomes intimately mixed with the mineral soil.

Since soil organic matter or humus takes up much more volume than the same weight of mineral particles, humification gave mineral soil a more open architecture with much more pore space. One percent soil organic matter by weight is about five percent by volume, so a soil A horizon with two to three percent organic matter by weight, which are very common values, contain about ten to 15 percent organic matter by volume. In general, the more organic matter in an A horizon, the higher the infiltration rate. Soil surface layers with more organic matter not only take in more water, they also retain more of it and make more of it available to plant roots. With the evolution of A horizons, the water-storage and water-supplying capacity of the world's soils were at least doubled, greatly enhancing plant growth and biomass production. At any given time, half or more of the water in soil surfaces worldwide is due to the presence of organic matter.

Organic matter has a similar effect on soil chemistry. Humification that took place during the Devonian led to a near doubling of cation exchange capacity in soils. As detailed in an earlier chapter, the cation exchange capacity of soil humus is at least ten times that of an equivalent amount of clay. Because of this, half or more of the cation exchange capacity in modern soils is due to negative charges on the surfaces of organic matter. This, in turn, means that half or more of the plant-available calcium, magnesium, and potassium in the world's soils would not be there if not for the presence of organic matter.

We have described how Histosols began forming in wet, low-lying land areas sometime after 470 million years ago. Different scenarios played out on better drained sites, where more intense weathering and large amounts of water percolating down through soil profiles brought about the evolution of three additional soil orders: Spodosols, Alfisols, and Ultisols. Unless plowed or otherwise disturbed, soils in all three of these orders exhibit the A-E-B profile sequence characteristic of modern soils. Accordingly, pedologists commonly refer to them as the "well differentiated forest soils." Spodosols are distinguished from both Alfisols and Ultisols by the presence of a spodic (Bh or Bs) horizon instead of an argillic (Bt) horizon. Both spodic and argillic horizons form as a result of percolating water picking up material in the upper part of the soil profile and then redepositing it farther down, but the material being translocated is different in Spodosols than in Ultisols and Alfisols.

Spodic horizons and thus Spodosols are formed by the translocation of iron, aluminum, and organic compounds from the upper part of the soil profile to the subsoil. Spodosols are an especially interesting class of soils and have been studied extensively for more than a century. Many early pedologists were Russian or European, and there are a lot of Spodosols in the glaciated landscapes of both Russia and northern Europe. Spodosols are striking in appearance, their most notable feature being the prominent gray or white E or eluvial horizon just below the surface. The E horizon is created when iron, aluminum and organic compounds are stripped away and then deposited at a lower depth, leaving the uncoated mineral grains behind. This soil-forming process is called podzolization. To see good examples of Spodosols, readers are encouraged to view the profile pictures of Marlow, the state soil of New Hampshire and Turnbridge, the state soil of Vermont, on the NRCS State Soil website.

Argillic horizons, the defining feature of Alfisols and Ultisols, evolved during the Devonian Period because of increased clay production and greater leaching intensity. Rain water percolating down through the soil profile picks up clay particles on the way. At some point, the water is immobilized by small pores or taken up by plant roots and the clay is redeposited. This process, referred to as clay translocation or by the French term lessivage, is still operative today. Only a little clay is translocated with each rain event, but as A and E horizons continue to lose clay a little bit at a time over many years, they become progressively coarser in texture. At the same time, the clay content of the underlying Bt horizon continues to increase. As the process goes on decade after decade, the textural difference between the eluvial (A and E) and illuvial (Bt) horizons becomes ever more pronounced and the boundary between them more abrupt. As more clay is removed from the surface layers, porosity increases, further increasing the infiltration rate and leading to even more clay translocation. As detailed in an earlier chapter, argillic horizons take a long time to form, so the stable landscapes that evolved during the Devonian Period also were a factor. To see a good example of Ultisols, readers are encouraged to view the profile picture of Bama, the state soil of Alabama, on the NRCS State Soil website.

Only a few well preserved soils dating from 450 to 350 million years ago have been found. McPherson (1979) described a buried Alfisol that was more than 360 million years old, while Kidston and Lang (1921) found a Histosol in Scotland that was nearly 400 million years old. Gillespie et al. (1981) later found a buried Histosol of similar age in West Virginia. The paleosol record for early Spodosols and Ultisols is more limited. There are as yet no well-documented paleosols of either order much older than about 30 million years, although Percival (1986) did find a buried soil more than 300 million years old that nearly met the taxonomic requirements for Spodosols. We join Retallack in lamenting the "present paltry record of forested paleosols." Although direct evidence in the form of buried soils is limited at this time, it is highly probable that, as stated by Berner (1997), "the first well-differentiated forest soils appeared in the Devonian." In other words, Spodosols, Alfisols, and Ultisols first evolved sometime around 400 million years ago and were widespread by 360 million years ago.

The Seed Bank

Soil taxonomists tend to focus on the effects of translocated clay on the argillic horizon, but the removal of clay from the overlying soil surface is far more important from an ecological and agricultural standpoint. The partial removal of clay from soil surfaces, along with the addition of humus, accounts for their characteristic loose structure and high porosity. The development of A horizons during the Devonian was fortuitous because of something else that was going on at the same time — the evolution of seed plants. Mosses and ferns, the earliest land plants, reproduced by releasing spores into the surrounding environment. Some of these spores grew into vegetative structures that produced sperm, while others grew into vegetative structures that produced eggs. Both kinds of vegetative structures are referred to as gametophytes. Upon maturing, sperm cells were released into the environment. They then had to swim to awaiting eggs and fertilize them, but water had to be present for this to happen. Because of the need for at least a film of water, plants such as mosses and ferns were largely restricted to swamps, seeps, or similar areas that were wet during some part of the year.

Seeds, which began evolving during the Devonian, were a big improvement. In seed plants, water is not needed for fertilization; it can take place by way of wind, insects, or even birds. The resulting embryo, along with a temporary food supply, is then enclosed in a tough, protective coating; this is the seed. Because of its protective covering, a seed can lie dormant for a long time, sometimes even years, waiting for an opportune time to grow. But seeds need a safe place to bide their time while waiting to germinate, and the A horizons that evolved during the Devonian met that need. They provided a nurturing environment in which young fragile roots could emerge from the seed and begin taking in water, and where newly formed leaves could easily make their way up to the surface and break free into the sunlight.

The blanketing effect of soil A horizons also helped to moderate extreme site conditions that might otherwise have limited the rapid spread of seed plants across the Earth's surface. Millions of years later, A horizons would help make human agriculture possible by serving as seedbeds for food crops. Because of the loose structure and loamy texture created by eluviation and humification, the surface layers of most soils are tillable even with primitive tools. One cannot overemphasize this point: The mobilization and removal of clay from A and E horizons (eluviation) is of much greater significance to ecology and agriculture than its subsequent re-deposition in the subsoil (illuviation).

Trees and Clay Evolution

Chapter 4 emphasized the differences between 2:1 clay such as mont-morillonite and 1:1 clay such as kaolinite. Montmorillonite and kaolinite were used as examples because they are so common in soil. Montmorillonite is very common in Mollisols, while kaolinite is ubiquitous in Ultisols and Oxisols. But that has not always been the case; neither of these clays was common before the evolution of vascular plants. According to Retallack (1990), "Both kaolinite and smectite [of which montmorillonite is an example] are more common after Devonian time and rare before that." As detailed earlier, both weathering intensity and leaching intensity increased dramatically during the Devonian, and this would have increased the prevalence of kaolinite by quickly removing elements such as magnesium and silicon from the clay-forming environment.

Montmorillonite became more common after 400 million years ago for a different reason, and it has a lot to do with potassium. According to Black (1957), "Some plants and plant tissues accumulate rather large concentrations of potassium. Tobacco leaves, for example, may contain as much as 8 per cent potassium on [a] dry weight basis." Many tree species take up and store a lot of potassium in their tissues. As pointed out in Chapter 5, the name potassium comes from "potash," referring to wood ashes in a pot. This reflects the fact that, for many centuries, people concentrated potassium salts by burning wood and then mixing the ashes with lye and boiling the mixture in a pot. The end product was used for manufacturing such things as glass, soap, and gunpowder and for dyeing and bleaching fabrics.

Table 9.2. Average potassium content (percent by weight) of montmorillonite and illite clay samples

Calculated from data of Grim (1953)

Clay type	Percent potassium by weight
Montmorillonite	0.13 (n = 14)
Illite	3.89 (n = 11)

Before forests covered the land and began extracting large amounts of potassium from the root zone, high-potassium clays such as illite dominated the world's landscapes; but soon after deep-rooted trees evolved, low-potassium clays such as montmorillonite were favored. The appendix of Ralph Grim's 1953 text, *Clay Mineralogy*, includes total chemical analyses of clay samples from around the world. Table 9.2 below shows the average amount of potassium (percent by weight) for samples of montmorillonite and illite; data are from Grim's appendix tables C and D. These numbers support the idea that depleting the clay-forming environment of potassium, as has been happening since the Devonian, favors the formation of low-potassium clays such as montmorillonite over high-potassium clays such as illite (see Knoll and James 1987).

Grass and Soil

By 360 million years ago, eleven of the twelve soil orders were in place. Another 300 million years would go by before Mollisols, the last of the soil orders, would appear. This chapter has emphasized the crucial role that forests played in the formation of Spodosols, Alfisols, and Ultisols. Grass played an equally important role in the evolution of Mollisols. Grass pollen showed up in the fossil record shortly before the dinosaurs went extinct about 65 million years ago. By 50 million years ago, grasslands occupied much of the Earth's surface and Mollisols, a new soil order, had formed beneath many of them.

The defining feature of Mollisols (from Latin mollis, "soft") is a thick, dark surface layer which results from the mixing of large amounts of organic matter with the mineral soil surface. About a third or more of grass roots die off each year, resulting in a steady supply of new organic matter. The above ground parts of grass plants also die off each year, and they too are returned to the soil where they quickly decompose. Adding so much organic matter to the soil each year soon results in the thick, black surface layers so characteristic of grasslands.

Mollisols, among the most fertile and productive soils in the world, are widely used for agriculture. They occur mostly in the middle latitudes and are especially extensive in prairie regions such as the Great Plains of North America and the "black soil" regions of central Europe and Russia. America is especially blessed when it comes to these desirable soils; Mollisols occupy more than 20 percent of the nation's land area. As indicated earlier, the State Soils of Nebraska, Illinois, Iowa, Kansas, Montana, North Dakota, Oklahoma, and South Dakota are all Mollisols, evidence of just how extensive these soils are. Readers are encouraged to view the profile pictures of Harney, the state soil of Kansas or Holdrege, the state soil of Nebraska on the NRCS State Soils website. Both are good examples of

Mollisols, clearly showing the dark mollic surface that is definitive for the order. Drummer, the state soil of Illinois, also is a good example.

The Soil Orders Summarized

This section provides summary information for each of the soil orders, including the basis for their classification, their most distinguishing characteristics, when and why they first evolved, and their relative value for agricultural or other uses. The percentage of the world's ice-free land surfaces occupied by each order is estimated, along with similar data pertaining to the total land area of the United States.

Entisols were the earliest soils to evolve on Earth, first appearing at least 3.8 billion years ago. They show little evidence of soil formation except, in many cases, an A horizon darkened by organic matter. They are common on recent alluvium and on steep slopes where erosion and instability retard soil development. They also are common on parent materials that resist weathering. For example, extensive areas of Entisols occur on the deep surficial sands of both the Nebraska Sandhills and the Carolina–Georgia Sandhills. There are extensive areas of Entisols on the floodplains and deltas of the world's large river valleys, where they provide food for millions of people. Entisols are the most extensive soil order, occupying nearly 20 percent of the world's ice-free land surface and approximately twelve percent of the land area in the United States.

Inceptisols (from Latin *inceptum*, "beginning") evolved at about the same time as Entisols, well before 3.5 billion years ago. They are more developed than Entisols, but still lack the horizons or features that define the other soil orders. Like Entisols, they commonly form on steep, unstable slopes, on recent alluvium, or in resistant parent materials. Inceptisols occupy about 15 percent of the world's ice-free land surface and approximately ten percent of the land area in the United States. There are large areas of Inceptisols on floodplains and deltas of the world's major rivers. This soil order provides food for about one fifth of the world's population.

Andisols (from Japanese *ando*, "black soil") form in volcanic ash or other volcanic ejecta. Environmental conditions conducive to the formation of Andisols have existed on Earth for more than 3.0 billion years, so they are believed to be one of the earliest soil orders to evolve. Andisols differ from other soils in that they are dominated by glass and colloidal weathering products such as allophane and imogolite. As a result, Andisols have unique chemical and physical characteristics that pedologists refer to as "andic properties." These include high water-holding capacity and the ability to "fix" large quantities of phosphorus and make it unavailable to plants. Andisols are the least extensive soil order, accounting for approximately one percent of the world's ice-free land area and a little under two percent of the land area of the United States, mostly in the Pacific Northwest.

Gelisols (from Latin *gelare*, "to freeze") are perennially frozen soils that form under very cold or periglacial conditions. They are limited to circumpolar regions and to high mountain elevations. Gelisols evolved shortly before 2.0 billion years ago as a consequence of the Earth's first major period of glaciation. Because of cold temperatures, Gelisols show little morphological development. Organic matter added to the soil decomposes very slowly, so Gelisols store large quantities of carbon. Because of the extreme environments in which they occur, Gelisols support less than 0.5 percent of the world population, the lowest percentage of any soil order. They occupy about nine percent of the land area of both the Earth and the United States. In the United States, they occur mostly in Alaska.

Vertisols (from Latin *verto*, "turn") are very clayey soils that shrink and crack when dry and then swell and expand upon rewetting. This shrinking and swelling action can cause serious engineering problems such as slumping roadways and cracking foundations. The repeated churning also prevents the soil from forming distinct, well-developed horizons. The first Vertisols evolved around two billion years ago in the interiors of small continents that had begun forming at that time. Vertisols now occupy about 2.5 percent of the world's ice-free land area and about two percent of the land area in the United States. The most extensive areas of Vertisols in the United States are in Texas.

Oxisols (from French *oxide*, "oxide," a compound of oxygen and another element) are highly weathered, naturally infertile soils that occur in tropical and subtropical regions. Agriculture is limited by low nutrient reserves, high phosphorus retention by iron oxides and low cation exchange capacity (CEC). Despite these limitations, many Oxisols have good physical properties and can be quite productive for agriculture with proper inputs of lime and fertilizer. The first Oxisols evolved shortly after two billion years ago as a consequence of a sharp rise in atmospheric oxygen sometimes referred to as the Great Oxidation Event. Oxisols now occupy about 7.5 percent of the world's ice-free land area. They do not occur in the continental United State but are present in Hawaii and in Puerto Rico.

Aridisols (from Latin *aridus*, "dry") occur in areas receiving very little rainfall and/or in areas where potential evapotranspiration greatly exceeds rainfall throughout the year. Areas in which Aridisols occur generally receive less than ten inches (25 centimeters) of rainfall yearly and most of that quickly evaporates. Since the soil is wetted infrequently and to only a shallow depth, leaching intensity is very low and there is little subsurface horizon development. Compounds such as sodium chloride, calcium sulfate, calcium carbonate, and silica commonly accumulate near the surface. Aridisols are used mainly as wildlife habitat, as rangeland, or for recreation. Agriculture is not feasible without irrigation. This soil order evolved between two billion and 1.5 billion years ago, coinciding with the appearance of the first deserts. Aridisols now occupy approximately twelve percent of the world's ice-free land area and a little over eight percent of the land area in the United States.

Histosols (from Greek *histos*, "tissue") are soils consisting primarily of decomposed organic matter. They have a layer at least 40 centimeters (16 inches) thick that contain at least 20 to 30 percent organic matter by weight. Bulk density is quite low, about one fifth the density of mineral soil. Histosols often are referred to as peats or mucks and have physical properties that restrict their use for engineering purposes. They typically form in low-lying areas where poor drainage slows the decomposition of organic matter, allowing it to build up on the surface. They are important ecologically because of the large quantities of carbon they contain. Histosols began to evolve shortly after 470 million years ago and probably were widespread by 400 million years ago. They now cover a little over one percent of the world's ice-free land area and about 1.5 percent of the United States.

Spodosols (from Greek *spodos*, "wood ash") are photogenic soils that typically have a white or gray eluvial horizon overlying a reddish-brown spodic (Bs or Bh) horizon. The spodic horizon results from the subsurface accumulation of iron and aluminum oxides, often complexed with humus. Spodosols commonly develop in coarse-textured parent material under coniferous forests in areas where rainfall is abundant, temperatures are cool, and leaching is intense. Because Spodosols are naturally acid and infertile, they require lime and fertilizer in order to be productive agriculturally. Spodosols first evolved around 400 million years ago and were widespread by the end of the Devonian, around 360 million years ago. They now occupy about four percent of the world's ice-free land area and about 3.5 percent of the land area in the United States.

Alfisols are moderately weathered, moderately leached soils that form primarily under temperate forests from base-rich parent materials. The argillic horizon, a subsoil layer enriched by clay washed in from above, is a defining feature of Alfisols. Because of the favorable climates under which they form and their relatively high native fertility, Alfisols are well suited to agriculture. The state soils of Ohio, Indiana, Kentucky, Minnesota, New York, and Wisconsin are all Alfisols, an indication of how important this soil order is to America's eastern farm belt. Alfisols first evolved around 400 million years ago and were widespread by the end of the Devonian, around 360 million years ago. They now occupy about ten percent of the world's ice-free land surface and about 14 percent of the land area in the United States. Alfisols support between 15 and 20 percent of the world's population.

Ultisols (from Latin *ultimus*, "last") are strongly leached, acid forest soils with relatively low native fertility. They occur primarily in humid temperate, subtropical, and tropical areas of the world, typically on old, stable landscapes. Because of intense weathering, much of the Ca, Mg, and other bases have been leached from these soils. Ultisols have distinctive subsoil (argillic) horizons in which illuvial clay has accumulated. Well drained Ultisols typically have yellowish or reddish subsoils indicating the presence of oxidized iron. The red clay soils of the southeastern United States are an example. Because of the favorable climates in which they occur, many areas of Ultisols are well suited to commercial forestry. Due to high acidity and low fertility, Ultisols are poorly suited to continuous agriculture unless limed and fertilized. Along with Alfisols and Spodosols, Ultisols first evolved around 400 million years ago and were widespread by the end of the Devonian, about 360 million years ago. They now occupy about eight percent of the world's ice-free land area. They are the dominant soils of the southeastern United States and account for about nine percent of the nation's total land area. Despite their acidity and low fertility, Ultisols are used extensively for agriculture and support nearly 20 percent of the world's population.

Mollisols (from Latin *mollis*, "soft") are the soils of grassland ecosystems. They are characterized by a thick, dark surface horizon that soil taxonomists refer to as a mollic epipedon. The mollic epipedon, which must be present in order for a soil to be classified as a Mollisol, results mostly from the long-term addition of organic matter derived from grass roots. Mollisols are among the most productive agricultural soils in the world and are extensively used for this purpose. They began to evolve around 65 million years ago, along with the worldwide expansion of grasslands that began to take place shortly after the demise of the dinosaurs. Mollisols occur mostly in the middle latitudes and are especially extensive in prairie regions such as the Great Plains of North America and the "black soil" areas of central Europe and Russia. They now occupy about seven percent of the world's ice-free land area and are the most extensive soil order in the continental United States, accounting for more than 20 percent of the total land area. The United States is especially endowed with these very productive soils, having three times the world average.

Pedologie and Agrologie

Friedrich Albert Fallou was the founder of German soil science. Trained as a lawyer, he worked with legal matters relating to the appraisal and taxation of land for many years. Because of a growing concern with what he perceived as the decline of soil quality in his region, Fallou eventually retired from the practice of law and devoted his time to the study of soil. In one of his many books, *Pedology or General and Special Soil Science*, written in 1862, he explained why the study of soil is so important and urged that pedology be established as a scientific discipline. Although eventually overshadowed by Dokuchaev in Russia and Hilgard in the United States, it could be argued that he is the true founder of soil science. In his 1862 book, Fallou took pains to distinguish between what he called *pédologie*, the scientific study of soil for its own sake, and *agrologie*, the practical study of agriculture, including the study of soil as it relates to farming.

This book thus far has focused mostly on *pédologie*, although some issues related to *agrologie* have crept in. The next chapter deals more explicitly with matters related to agriculture. This includes a consideration of why Jared Diamond (1987) famously referred to agriculture as "the worst mistake in the history of the human race." But most of the next chapter is devoted to discussing how the shortage of a single chemical element in the soil, nitrogen, limited agricultural production and, as a result, the size and health of human populations for thousands of years.

CHAPTER 10. NITROGEN: A PERVERSE CHEMISTRY

Sir William Crookes, in an 1898 address to the British Association for the Advancement of Science, offered the following grim assessment: "England and all civilized nations stand in deadly peril of not having enough to eat... we are drawing on the earth's capital, and our drafts will not perpetually be honored." He sounded this alarm because the rapidly growing populations of Europe and Asia were already using nearly all available land and still were having trouble feeding their people. It was clear that something needed to be done quickly or famine and social unrest were unavoidable.

Fortunately, the crisis Crookes warned against was averted. During the 20th century, especially after World War II, the United States and other industrialized nations struggled, not with food shortages, but with an excess of agricultural production. In addition to feeding rapidly growing populations, the industrial economies greatly increased the number of calories and the amount of protein available to each person. However, this did not silence the doomsayers, who argued that worldwide famine had not been averted, only postponed. Notable among them was Professor Paul R. Ehrlich of Stanford University, who in a 1968 book titled *The Population Bomb*, proclaimed that "the battle to feed all humanity is over." In an article the following year, Ehrlich warned, "Most of the people who are going to die in the greatest cataclysm in the history of man have already been born." Ehrlich's book sold over two million copies and was widely popular on college campuses, but his dire predictions, like those of Crookes 70 years earlier, failed to materialize.

To understand why the doomsayers were wrong, let us return to Crookes's 1898 speech, in which he hints at a solution to the problem, "It is the chemist who must come to the rescue. It is through the laboratory that starvation may ultimately be turned into plenty." Why would Crookes think that a chemist working in a laboratory could prevent worldwide famine? Very simply, he knew that the underlying problem with food production was a shortage of the essential element nitrogen. He also knew that the Earth's atmosphere held vast quantities of this nutrient. The problem lay in getting it out of the air and converting it into a form that crops could use, a question of chemistry. According to Crookes, solving the nitrogen problem and solving it soon was "vital to the progress of civilized humanity."

Nitrogen and Life

It is clear from the table below that humans are not fashioned from the clays of the earth; instead we are made up overwhelmingly of three elements taken from air (oxygen, carbon, and nitrogen) and one derived from water (hydrogen), and we are not unique in this respect. Every living thing on Earth consists mostly of these four elements.

Table 10.1. Elemental composition of the human body

Element	Percent of body weight	Origin
Oxygen	65	Air
Carbon	18	Air
Hydrogen	10	Water
Nitrogen	3	Air
Total	96	

Although present in much smaller quantities than either oxygen or carbon, nitrogen plays especially prominent roles in the makeup and functioning of living things. It has been said that carbon provides the quantity of life while nitrogen provides the quality. For example, most plant leaves appear green to the human eye because they contain chlorophyll, which reflects light in the green part of the spectrum. Chlorophyll, because it is vital to the process of photosynthesis, is one of the most important molecules on Earth. There are several different kinds, but the most common form in land plants is chlorophyll a, with the formula $C_{55}H_{72}O_5N_4Mg$. Note that chlorophyll a contains four nitrogen atoms. The other kinds of chlorophyll are b, c1, c2, d, and f, but regardless of type, at the heart of every molecule of chlorophyll on Earth are four nitrogen atoms surrounding a single magnesium atom.

In addition to being an essential part of the chlorophyll molecule, nitrogen is integral to the formation of proteins, which play such important roles in animal structure and metabolism. Proteins consist of folded chains made up of many smaller entities called amino acids. The term amino refers to the NH_2 or amino group that is a necessary part of every amino acid. In addition to the NH_2 group, each amino acid contains a carboxyl group (COOH) and an R group (or side chain); the R group is unique to each amino acid. There are 20 different amino acids and their sequence and the way the resulting chains are folded determine the structure and function of each protein. Listed below are some proteins that play important roles in the human body.

- Pepsin that helps to break down food in our stomachs

- Antibodies that help fight infections

- Proteins such as actin and myosin that enable muscles to contract and move

- Structural proteins such as collagen, elastin, and keratin that hold body tissues together

- Hemoglobin in the blood that transports oxygen throughout our bodies

- Insulin that regulates sugar metabolism and fat storage

- Cohesin protein, or "chromosome glue," that makes cell division possible

This list is just a small sample; scientists estimate that there are 20 to 25,000 genes that code for different proteins in the human body, and nitrogen is a necessary part of every amino acid in every protein. In order to function and stay alive, our bodies must retain about three percent nitrogen by weight at all times. This is a little over three pounds (1.4 kilograms) for a 120-pound woman and a little over five pounds (2.3 kilograms) for a 180-pound man. Because protein, along with the nitrogen it contains, is continuously lost from the body, it must be replenished on a regular basis, meaning that we need to consume protein on an almost daily basis. Getting the 75 grams or so of protein that adults need each day has been a problem throughout much of human history, and the underlying problem has been nitrogen, or rather the lack of it. As a result, much of the world's population has been nitrogen starved and protein starved for the last 8,000 years.

The problem is one of chemistry and more specifically, the perverse chemistry of nitrogen.

Nitrogen exists in the atmosphere as highly inert N_2 molecules held together by very strong triple bonds ($N{\equiv}N$). Nitrogen cannot be used to make chlorophyll or to build plant and animal protein until it is removed from the atmosphere and the bonds holding the $N{\equiv}N$ molecules together are broken apart, and this happens only two ways in nature. One is by way of lightning. Up to eight million bolts of lightning strike the earth each day and millions more zigzag across the sky without ever reaching the ground surface. Each flash produces unimaginable heat in only a few millionths of second. This immense surge of heat and energy breaks apart many of the $N{\equiv}N$ bonds. As the incredible heat is dissipated, freed nitrogen quickly combines with oxygen in the air to form nitric oxide (NO). The NO is short lived, combining on contact with another oxygen atom to form nitrogen dioxide (NO_2). The NO_2 readily dissolves in rainwater and is washed out of the air and deposited on the earth's surface. The dissolved nitrogen is carried down into the soil, where it can be taken up by plant roots. In what many consider a prelude to modern biogeochemistry, Jean-Baptiste-André Dumas concluded his course at the Sorbonne in 1841 with a lecture describing the role of lightning in nitrogen fixation (Aulie 1970).

> As it is from the mouths of volcanoes that the principal food of plants, carbonic acid, is incessantly poured out; so it is from the atmosphere on fire with lightnings...that the second and scarcely less indispensable aliment of plants, nitrate of ammonia, is showered down for their benefit... By the agency of light, carbonic acid yields up its carbon, water its hydrogen, nitrate of ammonia its nitrogen. These elements combine, organic matters are formed, and the earth is clothed with verdure.

Dumas assumed that lightning was the only source of plant-available nitrogen, unaware that while lightning is noisily splitting apart N_2 molecules in the air, select groups of microorganisms are silently doing the same thing underground. Nearly 50 years after Dumas gave his lecture, Hellriegel and Wilfarth (1888) published a description of how certain microbes can extract nitrogen (N_2) from the atmosphere and convert into a form that plants can use. Here are some highlights from their paper.

> The Gramineae [grasses] are solely dependent on the nitrogenous compounds present in the soil. To the Leguminosae a second source is available...the free elemental nitrogen in the air. The Leguminosae do not themselves possess the ability to assimilate the free nitrogen of the air, but the active participation of living microorganisms in the soil

is absolutely necessary. It is necessary that certain kinds of the latter [microbes] enter into a symbiotic relation with the former [legumes].

The microbes that form symbiotic relationships with legumes are among the smallest organisms on Earth. They start out as tiny, immobile spheres, but after changing their form and growing a tail, they move through the soil until they approach the root of a receptive plant. The root, upon "sensing" their presence, secretes a substance that causes the strange little creatures to swim toward it. The microbes then release a substance into the soil that causes the plant root to bend. The microbes enter the root at that spot, settle in and begin to multiply, soon forming a visible nodule. A partnership has been formed. The plant provides the microbes with shelter, water and food and in return, the little visitors take N_2 out of the air and change it into a form the plant can use to grow. Other kinds of nitrogen-fixing bacteria form partnerships with plants other than legumes. For example, red alder (*Alnus rubra*) is a prolific tree species that grows along the west coast of North America from California to the southern border of Alaska. It is able to colonize disturbed areas and grow rapidly because it establishes a symbiotic relationship with *Frankia*, a nitrogen-fixing actinomycete. Some free-living microbes also fix nitrogen; but the total number of nitrogen-fixing microbes in the world, whether symbiotic or free living, is rather small.

Any microorganism that can fix nitrogen is able to do so because it contains the enzyme nitrogenase, which is one of nature's nanomachines, ancient molecules that, according to Paul Falkowski (2015), "carry out the necessary functions of all living things." Nitrogenase performs the vital task of converting N_2 to NH_3, which plants can then use to make chlorophyll and proteins. Nitrogen in the N_2 form is useless to them. Nitrogenase, like the other nanomachines, evolved very early in Earth's history, and although it is widely distributed, there is a dismayingly small amount of it in the world; according to Smil (1997), "the total mass of this unique enzyme may be less than a dozen kilograms." This is about 25 pounds for the whole world.

Here is another interesting fact about nitrogenase. If you pull a legume out of the ground during the growing season and slice one of the root nodules open, you will see that the cut surface has a pinkish tinge, indicating the presence of leghemoglobin. This plant substance is very similar to the hemoglobin found in human blood and is necessary for the nitrogen-fixing process to work efficiently. When nitrogenase first evolved, the world was devoid of free oxygen. As the atmosphere became increasingly oxygenated, plants and microbes developed several strategies to keep nitrogenase from being degraded by oxygen. For example, cyanobacteria in the oceans fix nitrogen in special protective structures called heterocysts, but this strategy is only partly successful; up to a third of all nitrogenase in ocean environments is deactivated by oxygen before it can fix very much nitrogen. On land, the nitrogenase in root nodules seems to be better protected; leghemoglobin, with its high affinity for oxygen, is fairly effective at safeguarding the nitrogenase.

There is yet a third issue with nitrogenase; it uses up a lot of biological energy. According to Raymond et al. (2004), "the nitrogenous enzyme systems, at 16 ATP's hydrolyzed per N_2 fixed, carries out one of the most metabolically expensive processes in biology." But despite its metabolic cost, nitrogen fixation is well worth it to the plants involved, because nitrogen has such a pronounced effect on growth and reproductive success. Plants that take in more nitrogen are able to create more leaf surface and produce more light-capturing chlorophyll. This enables the plant to carry out photosynthesis at an enhanced rate, producing more food (glucose) and more energy (ATP molecules).

Every year lightning and microbes enrich the world's soil with nitrogen, making it possible for terrestrial plants and animals to thrive. The thrifty Earth builds up a reserve supply by storing some of this nitrogen in organic matter at the soil surface. The amount of nitrogen supplied by microbial fixation each year varies with such things as temperature, rainfall, and the number of nitrogen-fixing plants present. It ranges from only a few pounds per acre to 100 pounds (45 kilograms) or more in *Acacia* savannas or in forests dominated by tropical legumes. On most sites, microbes are believed to fix more nitrogen than lightning; common estimates of lightning-derived nitrogen range from five to 15 pounds (2.3 to 6.8 kilograms) per acre.

Imagine an area of land on which microbes add 40 pounds (18 kilograms) of nitrogen per acre to the soil each year, while lightning adds an additional ten pounds (4.5 kilograms), for a total of 50 pounds (22.7 kilograms) per acre. In addition to the 50 pounds provided by lightning and microbes, we can add the nitrogen coming from a third source. Over time, nitrogen builds up in the soil as part of the organic matter, and each year a small part of this is made available for plant growth. In temperate zones, this begins to happen in the spring when soil fungi and bacteria wake up from their winter sleep and start breaking down organic matter for energy, releasing plant available nitrogen in the process. If we know the soil organic matter content at a given location, we can estimate the amount of stored nitrogen that will become available each year.

Assume that the A horizon at a hypothetical site is 15 cm (6 inches) thick and contains three percent organic matter by weight. Agronomists tell us that a soil layer this thick typically weighs about two million pounds. Soil weight times the percent organic matter (2,000,000 pounds x 0.03) gives us the amount of organic matter per acre, which in this case is 60,000 pounds. On average, soil organic matter contains about five percent nitrogen by weight. Multiplying the weight of organic matter per acre (60,000 pounds) by 0.05 gives us the amount of nitrogen per acre, 3,000 pounds. Finally, agronomists estimate that one to two percent of the nitrogen in soil organic matter becomes available each year. Multiplying 3,000 pounds by 0.01 and then by 0.02 gives us the approximate range of plant available nitrogen released by the breakdown of organic matter by microbes each year; the answer is 30 to 60 pounds per acre.

All three nitrogen sources vary across the land surface in response to a wide array of environmental factors. Lightning is beyond human control, but soon after the adoption of agriculture, farmers all over the world learned to manipulate the other two. Although they knew nothing of nitrogen, microbes, or legumes, early farmers observed that growing certain kinds of plants made the soil more productive. They soon identified a number of such plants and began to grow them as part of their farming practice. Early farmers all over the world also figured out that crops grew better in soils that contained more organic matter. They did not know the reason for this, but returning as much organic matter to the soil as possible soon became the goal of every sensible farmer in the world. But we are getting a little ahead of the story. Before delving too deeply into the cultivation of legumes or the recycling of organic matter, we need to address a more fundamental question — why, about 10,000 years ago, did humans all over the world begin to give up hunting and gathering to become farmers?

The Hunter-Gatherers

The ancestors of modern humans roamed the African savannas for many thousands of years and during that long time period, numerous species arose, flourished for a time, and then died out. Sometime before two million years ago, an especially clever species, *Homo erectus* or "upright man" appeared. *Homo erectus* had developed the ability to fashion hard stones into tools and weapons with cutting edges. Their efforts were clumsy at first, but by about 1.5 million years ago, they were producing finely made weapons and tools with sharp, finely flaked edges; these tools enabled early hunters to remove the tough skin from large animals and to sever muscle from bone. Stone tools also were effective in breaking open the skull and long bones of animals, providing access to the rich stores of fat in the brain and bone marrow. With sharp-edged stone weapons and tools, humans were able to kill, skin and butcher large prey, giving them greater access to life-enhancing fat and protein.

Homo erectus roamed the Earth for two million years or more before going extinct about 200,000 years ago, near the same time that modern humans, *Homo sapiens* or "wise man" arose. Scientists have analyzed human bones unearthed at two sites in Ethiopia and found them to be 195,000 and 160,000 years old, so we know that members of our species have been on Earth for at least that long. *Homo sapiens* benefited from having a brain about one third larger than *Homo erectus*. Despite occasional setbacks, *Homo sapiens* thrived and soon spread throughout much of the African continent. Around 60,000 years ago some of them migrated out of Africa and began to settle the Middle East; then about 45,000 years ago another migration began. Some of our ancestors headed to the north and east toward the Indian subcontinent and Asia. Others headed to the north and west, eventually settling Europe and what would eventually become the British Isles. By about 15,000 years ago, *Homo sapiens* had spread to every habitable continent and was the only hominid species left on earth.

Modern anthropologists have studied a number of hunter-gatherer groups that survived into the 20th century and have been amazed at the amount of environmental information these "primitive" people carried around in their heads. Hunter-gatherers, regardless of where they lived, had an almost encyclopedic knowledge of the plants and animals that provided them with food. One observer (Thomas 1959) noted that the Bushmen of southern Africa knew the life habits of more than 50 animals. They also knew the location of nearly every stone and bush in their territory and could take a visitor to every site of edible plants, fruits or nuts and describe the time of the year in which the foods would be available. They usually had names for such places, even if they were extremely small. Jared Diamond, in his influential book *Guns, Germs, and Steel*, described how the natives of New Guinea with whom he traveled could identify more than 20 species of local mushrooms. The Indians of North America were renowned for their detailed knowledge of plant and animal ecology. Tribes living in the Great Lakes region used more than 100 species of plants for food and more than 15 for flavoring foods and beverages.

By about 30,000 years ago, humans had developed an impressive array of tools and weapons. Their hunting arsenal included stone axes with keen cutting edges and sharp spear points fashioned from bones and antlers.

Groups living near bodies of water had learned to make fish traps, as well as fish hooks and nets. In total, the human toolkit included more than 100 items. Humanity also had mastered fire; they could summon it at will by generating sparks from flint and other stones. In addition, they soon learned to use it as a land management tool, setting fires regularly to make their hunting areas more accessible and to improve conditions for grazing animals. A small, unheralded invention, the eyed needle, was perhaps as important as fire in allowing humans to populate the colder regions of the earth. The first needles probably were made by sharpening a sliver of bone or ivory and then boring a hole in one end with a small flint awl. With needles, humans could sew multilayered cold weather clothing. This simple implement, along with fire, enabled tropical humans to adapt to almost any climate on Earth, including the frozen arctic.

In addition to a variety of weapons, tools, clothing, and the ability to make fire, humans living 30,000 years ago had developed an impressive food technology. They knew how to remove the bitter, sometimes toxic compounds present in many high-energy foods. Examples are acorns and legumes in temperate regions and cassava in the tropics. A common technique was to place acorns, legumes or other foods in a loosely woven bag and immerse them in a swiftly running stream for several days, allowing the flowing water to leach out the offending chemicals. The ability to remove the bitter tannins from acorns was especially significant, since oak forests are so prevalent in parts of the Middle East, Europe, and North America. Oak forests became a major source of food for many hunter-gatherer groups outside the tropics. Colonial records show that cooked acorns were one of the first foods offered by local Indians to the Puritan settlers of New England (Philbrick 2006).

Many people, when they think of hunter-gatherers, picture forlorn groups wandering aimlessly across the land, perpetually lost and not sure if they will find food the next day. Nothing could be further from the truth. Except for the few that followed migratory herds, all such groups had home bases from which they exploited a known territory. From a central camp, they could utilize the resources of more than 100 square kilometers simply by walking six kilometers (about four miles) in any direction. Most people can walk this distance in a little over an hour. The geometry of hunting and gathering is simple; the area of a circle equals pi, or 3.14, times the radius squared (r^2); in this case, that is 3.14 x $(6)^2$ or (3.14 x 36) = 113 square kilometers. If a hunter-gatherer band consisted of 25 people, a working circle this large would provide each member with about four square kilometers of food-producing area; this is 400 hectares or about 1,000 acres per person.

Observations of modern hunter-gatherers have shown that, in most environments, an area only half this size is needed. Many bands had two base camps — a summer and winter camp or a dry season and wet season camp. By moving just four or five miles in any direction, they had an entirely new range to exploit.

Hunter-gatherers had a wide variety of foods to choose from. About 30,000 plant species on Earth have some part, such as a fruit, a nut, a fleshy root, or a leaf that people can eat. It is estimated that humans have used about 7,000 plant species for food sometime in the past and about 150 of them became important crops; but many former food crops have now been abandoned. Just 15 plant species now provide more than 90 percent of our food, and just three — wheat, corn, and rice, provide two thirds of all our food energy worldwide (Paoletti et al. 1992).

Many anthropologists consider the time from 35,000 years ago until the adoption of agriculture to have been the golden age of human nutrition. Foraging and hunting provided a dependable, nutritious diet and a good life in general. Nearly everyone could easily obtain sufficient calories and the natural foods they ate were high in protein, vitamins and minerals. Early humans had little trouble feeding themselves and had enduring social networks, living all their lives in small bands consisting of relatives and companions they had known intimately from childhood. Jean-Jacques Rousseau described human existence before the advent of agriculture as a time when "men were strong of limb, fleet of foot and clear of eye...troubled with hardly any disorders save wounds and old age" (Williams 2014).

But until very recently, few anthropologists shared this viewpoint. Instead, most of them agreed with Thomas Hobbes, who in 1651 offered a harsh assessment of hunter-gatherer life (Malcom 2012): "No arts, no letters, no society, and, which is worst of all, continual fear and danger of violent death, and the life of man solitary, poor, nasty, brutish, and short." If anything, many of those engaged in hunter-gatherer studies during the first half of the 20th century strove to be more Hobbesian than Hobbes. R.J. Braidwood wrote a very influential anthropology textbook titled *Prehistoric Men* in 1948. This book went through six editions, the last one in 1964. Braidwood took a grim view of hunter-gatherer life, declaring that it was "all in all, a savage's existence, and a very tough one. A man who spends his whole life following animals just to kill them to eat, or moving from one berry patch to another, is really living just like an animal himself."

A seventh edition of Braidwood's book was never published, partly because of revelations that came out of a scientific meeting held at the University of Chicago in 1966. At this symposium on Man the Hunter,

America's leading anthropologists gathered to discuss contemporary hunter-gatherers and their mode of life. Many of the attending scientists had actually lived with hunter-gatherers and learned what their lives were really like, and the truth was nothing like the conventional wisdom that had been espoused for so many years. Lee and Devore (1968), who carried out some much-cited studies of the !Kung people in southern Africa, captured the essence of this meeting with the following statement: "To date, the hunting way of life has been the most successful and persistent adaptation man has ever achieved."

Considering that there has never been any empirical evidence to support the idea that hunter-gatherers lived a forlorn, half-starved existence, why were early anthropologists so eager to believe it? Wharton (2001) gave what is arguably the best answer: "In order to believe that our society has 'progressed', we must believe first that the lives of our ancestors were indeed nasty, brutish and short." She continues, "But, as study after study has confirmed, the health of traditional peoples was vastly superior, in almost every way, to that of modern industrial man."

The Worst Mistake in History?

But regardless of how efficient they were at hunting and gathering and how good their lives were, only so many hunter-gatherers could live in a given area, typically only one or two per square kilometer, or about one person for every 100 to 200 hectares (250 to 500 acres). Hunter-gatherers, like humans today, probably took every opportunity to make their lives easier. One way was to reduce the amount of territory they had to cover by encouraging the growth of food plants near their camps. Foraging humans had an almost encyclopedic knowledge of everything growing in their territory and understood the conditions under which various plants grew best, so it was easy to make the transition from gathering food to growing food. In addition to those deliberately planted, some food plants probably invaded human camps on their own, taking advantage of disturbed sites such as pathways or old dwelling sites.

But creating gardens of food plants near their camps was not a perfect solution; with time and repeated harvesting, the organic matter in these garden soils would become depleted, reducing the amount of available nitrogen. Of course, early humans knew nothing of nitrogen, but they were astute enough to realize that after a few years, food production in an area of land often would decline. Being experts at pattern recognition, they undoubtedly noted the connection between disappearing organic matter and loss of productivity. Accordingly, they soon learned to exploit an area for a few years until the organic matter was depleted, then "shift" to another piece of land and let the original area return to woodland or grass. After some years they would return to the original plot and grow food on it again. In effect, they were systematically working their way across the landscape and harvesting the nitrogen stored in soil organic matter. In most areas, shifting agriculture would easily feed more than ten people per square kilometer, a carrying capacity five to more than ten times greater than hunting and gathering. Shifting agriculture was practiced for a long time in some parts of the world, but in the more productive areas, it soon gave way to settled or sedentary farming.

Twelve thousand years ago, everyone on Earth hunted and gathered wild foods, but by 8,000 years ago, nearly all of the human race had given up hunting and gathering to take up farming. Why did people all over the world abandon a subsistence strategy that had served them so well for so long? Jared Diamond (1987) famously referred to agriculture as the "worst mistake in the history of the human race" and French philosopher Jean-Jacques Rousseau (1755) had expressed similar views more than two centuries earlier.

> The first man who, having enclosed a piece of ground...saying "This is mine," and found people simple enough to believe him, was the real founder of civil society. From how many crimes, wars and murders, from how many horrors and misfortunes might not any one have saved mankind, by pulling up the stakes, or filling up the ditch, and crying to his fellows, "Beware of listening to this impostor; you are undone if you once forget that the fruits of the earth belong to us all, and the earth itself to nobody."

You might think that agriculture was a deliberate, conscious decision, but it was not. The so-called agricultural revolution was not the result of a choice or decision, and it was an evolutionary rather than a revolutionary process. In most early societies, hunting and gathering coexisted with limited agriculture for hundreds, sometimes even thousands of years. Succeeding generations were unaware that, slowly but surely, they were abandoning their hunter-gatherer existence and becoming farmers (Cohen 1977). Two well-documented examples of this come from the Middle East, one dealing with the cultivation of grains and the other with the domestication of animals.

Archaeologists studied the agricultural history of an early village in southwestern Iran by examining old refuse dumps layer by layer. About 9,000 years ago cultivated plants accounted for only five percent of the seeds recovered. The remaining 95 percent were seeds that had been gath-ered from wild stands. Within 1,000 years, the percentage of cultivated seeds had increased significantly but still represented only 40 percent of the total. This illustrates the normal course of events; early societies adopted agriculture not in a transforming flash, but very slowly over many hundreds of years (Budiansky 1999).

Excavation of an ancient settlement in Syria showed that animal agri-culture followed a similar pattern of long-term incremental change. The bones of domesticated sheep and goats first appeared about 9,500 years ago, when they accounted for about ten percent of the bones found at the site. The bones of wild gazelles made up most of the remaining 90 percent. Note: Archaeologists can tell whether bones came from a wild or domestic animal by the characteristic ratios of certain elements and by certain changes in the bone structure of domesticated animals. Gazelles were once numerous in the Middle East and the large migrating herds were a major source of food for early inhabitants (Legge and Rowley-Conwy 1987). As in the case of plants, animal domestication increased at a very slow rate. It took about 1,000 years for the bones of domestic sheep and goats to make up more than half of the total.

These examples show that agriculture did not suddenly replace hunting and gathering in the Middle East; instead the transition was slow and incremental, with agriculture playing only a supplemental role for many centuries. It is likely that the transition to agriculture followed a similar pattern in the other major world centers, such as China, Mexico, and Central and South America.

Recycling Organic Matter

As populations grew, farmers were forced to stay on the same piece of land for generations, but in order to do this they had to keep replenishing the soil's nitrogen. One strategy was to recycle animal waste and decaying plant parts. This soon became standard practice throughout the world. In the Odyssey, written around 3,000 years ago, Odysseus returns home and sees his dog Argus, "lying neglected on the heaps of mule and cow dung that lay in front of the stable doors till the men should come and draw it away to manure the great close." Recycling animal and plant waste was widely practiced during early Roman times. Marcus Cato, around 200 BC, advised Roman farmers to "spread pigeon dung on meadow, garden and field crops...Save carefully goat, sheep, cattle, and all other dung...You may make compost of straw, lupines, chaff, bean stalks, husks, and ilex and oak leaves."

Sung (1637) gave similar advice to Chinese farmers, telling them that they should recycle "human and animal excretions...grass and tree leaves." He made special mention of soybeans as a soil amendment: "When the price of beans is low, soybeans can be cast into the field, each bean enriching an area of about three inches square; the cost is later twice repaid by the grain yield." The Chinese, and later the Japanese, took Sung's advice about enriching the soil with human waste to heart. As populations increased in rice-growing areas of Asia, the recycling of human waste became widespread. In traditional Chinese farming, nearly 80 percent of all human waste from towns and cities was carried to the countryside and put on the soil. This custom was followed in some parts of China until the early 1980s. The Japanese followed similar practices. In 1649, authorities in the city of Edo, now Tokyo, ordered that all sewage conduits in the city be disconnected, and soon Japanese farmers were carrying away the municipal waste and spreading it on cropland. Many homes in Tokyo were not reconnected to a sewage system until well after World War II (King 1927, Tanaka 1998).

Legumes

Economies and cultures specific to certain food crops soon developed in various parts of the world. Middle Eastern agriculture was based mostly on wheat, while corn became the main crop in Mexico and Central America. Chinese agriculture began with millet production on the rich loess soils of the northeast. This was followed a few thousand years later by intensive, irrigated rice farming along major rivers. Interestingly, early rice farmers in China, wheat farmers in the Middle East, and corn farmers in Mesoamerica had one thing in common; all planted legumes in addition to their staple grain. Archaeologists have identified peas and lentils in the ruins of Middle Eastern agricultural sites dated as early as 10,000 BC. In mild climates, growing legumes during the winter could produce enough nitrogen for a good grain crop the following summer. Not surprisingly, all early agricultural systems made use of legumes such as beans, peas, soybeans, and peanuts. Before the 20th century, it was not possible to practice intensive agriculture and feed large populations without doing so.

Many early Roman writers were enthusiastic about legumes, sometimes offering farmers very specific advice. Theophrastus, around 300 BC, observed that "the bean best invigorates the ground...wherefore the people of Macedonia and Thessaly turn over the ground when it is in flower." In contrast, Columella favored alfalfa, arguing that it was "outstanding for several reasons...one seeding affords, for all of ten years thereafter, four harvestings regularly and sometimes six; it improves the soil; lean cattle of every kind grow fat on it." Modern data support Columella's observations. Alfalfa does seem to stand out as a nitrogen fixer; a productive field can fix 200 pounds or more of nitrogen per acre (224 kilograms per hectare), much more than soybeans, lentils or common beans (Havlin et al. 2016).

Farming Populations

Local populations typically increased ten or 20 times within only a few hundred years of adopting agriculture. Farming mothers did not breastfeed their babies for as many years as hunter-gatherers, weaning them early on gruel or porridge made from grain or soft starchy vegetables. This led to more frequent births and the result, in most cases, was explosive population growth. The early farming population of the Middle East is believed to have increased from fewer than 100,000 to more than three million people in only 160 generations. Other centers of agriculture throughout the world experienced similar surges in population, but most of the people in these growing populations were not happy and well fed (Farb 1978).

There was a curious and pernicious contradiction to agriculture. It would sustain a much larger population than hunting and gathering and could even provide a surplus in good years. But instead of rescuing humanity from hunger, agriculture actually made periodic famine more likely. It simplified the ecological system, so that populations came to rely on one crop or only a few crops for their total sustenance. Planting large areas to a single plant species made outbreaks of plant diseases or attacks by insects more likely as well as more devastating. As agricultural communities became larger and more complex, any breakdown of social order could result in food shortages. War with a neighboring people or a severe disease epidemic could lead to neglected crops or to crops being destroyed or stolen. Food would become scarce and the weak and the vulnerable would begin to starve.

Some parts of both Western Europe and Africa have, on average, experienced famine about every third year during the last 2,000 years. France, which had a more productive agriculture than most countries, experienced a total of 20 nationwide famines during the 15th and 16th centuries. China has been even more prone to hunger; until just recently, some part of the country had experienced severe famine every year for the past 2,000 years (Farb 1978). Contrary to what was once believed, agriculture did not rescue hapless hunter-gatherers from periodic starvation. Instead of being abolished, food insecurity and periodic famine soon became the defining characteristics of agriculture.

Under agriculture, hunger due to natural causes soon was made even worse by the rise of feudalism, an oppressive economic system that developed in all grain-producing regions of the world. Under feudalism, most of the world's farmers became peasants or serfs and were deliberately deprived of food, especially protein, by those who had assumed power over them. Persistent, hard labor at bare subsistence wages became the lot of most humans only a few thousand years after the adoption of agriculture. Most of the people now living on Earth, regardless of the country in which they live, are descended from such peasants. Exceptions include the descendants of hereditary elites such as ancient royal families and isolated groups of subsistence farmers and hunter-gatherers that survived into modern times. For generations, most of our ancestors were at the mercy of whoever owned the land on which they lived. In many cases a ruler had bequeathed it to some favored subject in the past as payment for political support or military service, and the land had since been handed down from generation to generation. Peasants typically came with the land and often were forbidden by law from leaving it. They were caught in a trap; a wealthy landlord, the nobility, or the church took all the products of their labor beyond those needed for the most meager subsistence.

The economic lot of peasants the world over is illustrated by the records of a small farm in the Mecklenburg region of northern Germany during the 1400s. This farm produced a little more than 10,000 pounds of grain annually. About 3,200 pounds of this had to be held over to plant next year's crop. Another 2,800 pounds was kept back to feed plow horses during the winter. This left 4,000 pounds of grain, enough to feed a family of four with plenty left over. But 2,700 pounds of this was appropriated by the landlord as rent. This left only 1,300 pounds of grain to feed the farmer, his wife, and their two children, or about 1,600 calories per person, not quite enough to keep laboring people in northern Europe alive. Families living under these conditions were forced to get additional calories by gathering foods such as roots, mushrooms, nuts and berries in the forest whenever such activities were possible. Other practices, dangerous ones, were sometimes employed, including hiding grain from the landlord or poaching his fish and game. Over time, the ruling groups of every feudal society learned how to leave the peasant just enough for the most meager survival (Wolf 1966).

The overriding injustice of peasant life was that, despite producing a surplus on the land they farmed, succeeding generations were forced to suffer from life-long calorie and protein deficiency. In the words of anthropologist Peter Farb (1978), "The peasant diet is unique among the basic human adaptations in that it contains almost no animal protein." In the example above, with only 1,300 pounds of wheat to feed his family for a year, the farmer could not afford to process it through an animal and then eat the meat; too many calories would be lost. A proclamation by King Henry IV of France upon ascending the throne in 1589 is particularly telling: "If God grants me the usual length of life, I hope to make France so prosperous that every peasant will have a chicken in his pot on Sunday." King Henry reigned more than 20 years and embroiled the country in a succession of expensive wars, while at the same time maintaining a lavish lifestyle at his court. But for some unexplained reason he never was able to improve the diet of his peasants.

Toward the end of the Middle Ages, it became increasingly clear that agricultural systems based on hungry peasant labor were unsustainable. But how would it all end? What would the answer be to the question posed by Edward Markham in his famous poem, *The Man with the Hoe* ?

> How will it be with kingdoms and with kings
> With those who shaped him to the thing he is
> When this dumb Terror shall rise to judge the world
> After the silence of the centuries?

Fortunately, most of the bloodletting feared by so many never came to pass. The French Revolution was a notable exception, and considering what happened there, the rather peaceful way peasantry ended in England is instructive. Beginning in the 1600s, agricultural land became more productive, largely due to the introduction of new crops such as corn and potatoes from the Americas and by the fact that landowners had optimized the use of legumes, crop rotation, and organic matter recycling. Instead of many small plots tended by peasants, English landowners wanted to cultivate larger blocks of land and use paid laborers. This meant that many peasants had to leave the land. The English parliament helped bring this about by allowing enclosures, a process by which landlords could void the traditional "rights" of peasants, evict them from the land, and consolidate the small tracts they once inhabited into large, enclosed landholdings. Oliver Goldsmith, in his poem *The Deserted Village*, describes a decaying community left behind after peasants were forced off their small plots by land consolidation. At the time the poem was published in 1770, at least half the population of England had already left the countryside and moved into towns to become wage earners (Christian 2005).

This rural exodus became more pronounced during the last few decades of the 1700s. This was largely due to the coming of the Industrial Revolution, thanks largely to James Watt's development of an efficient steam engine. In addition to locomotives and steamships, engineers quickly developed steam-driven machines for a number of industrial uses, and soon "dark satanic mills" dotted the English countryside. Peasants left the farm areas in droves, flocking to the cities and towns in search of work, fleeing what Karl Marx later called the "idiocy of rural life."

The peasants of Europe soon became the urban poor of Europe, and their living conditions improved not at all. As peasants moved to the cities, their meager peasant diets moved with them and they continued to work for starvation wages. Bread and to some extent potatoes became the dietary mainstays of the urban poor, both in northern Europe and in America. An 1841 survey of working class families in England found that they got half of their total calories from bread and another 20 percent from potatoes. Fifty years later, in 1892, another survey found that more than 80 percent of the children in one English community had eaten nothing but bread for 17 of their last 21 meals. As late as 1910, bread and potatoes still accounted for nearly half of the calories consumed by urban dwellers in both the United States and England. On the eve of World War II, the poor of England still ate mostly bread, with each person consuming about a pound and a half daily (Mount 1975).

At the beginning of the 20th century, most of the world's people remained poor and even in richer industrialized countries, ate near vegetarian diets consisting mostly of wheat, corn, rice, potatoes, and beans, with very little animal protein. On a worldwide basis, agricultural production barely met human needs, so nearly all of the grain produced had to be consumed directly by humans, with only about ten percent fed to cattle, hogs or other meat producing animals. In 1900, the world's farmers were feeding 1,625 million people a largely vegetarian diet by cultivating 850 million hectares of land (Smil 2001a). This means that, on average, each hectare of farmland was feeding about two people (1,625 divided by 850 = 1.9). Some farmers in parts of China and western Europe did much better, producing enough food for five or more people on each hectare; but the worldwide average was about two.

The Ages of Food

In terms of food, we can divide the history of Homo sapiens into three distinct ages. First came the age of the hunter-gatherer, which lasted from about 200,000 to about 10,000 years ago. During this time, humans lived much as wild animals still do, gathering and hunting the plant and animal foods provided by nature. The overall numbers and population density of hunter-gatherers remained low. Worldwide, the total population was never more than a few million, because each person needed 50 to 100 hectares of land, up to a square kilometer or more, to meet his or her yearly food needs.

Next came the age of organic or nonindustrial agriculture, which began about 10,000 years ago and lasted until early in the 20th century. During this time, humans cultivated a small number of plant species while using legumes and intensively recycling organic material to maintain soil fertility. By growing legumes and returning organic matter to the soil, farmers actually were "domesticating" and managing microbes, both those that take nitrogen out of the air and those that free up nitrogen by breaking down organic matter in the soil. Legume management and recycling methods improved with each generation, and human numbers grew steadily, reaching a worldwide population of about 1.6 billion by 1900.

But shortly after 1900, traditional organic agriculture gave way to industrial agriculture in much of the world. This was truly a revolution. We can cite the precise year, the precise day, and the precise event that signaled this change. The date was July 2, 1909, and the event was a laboratory experiment carried out by a German chemist named Fritz Haber. Haber described his historic accomplishment in a letter sent to the directors of his company on July 3, 1909, writing in very understated scientific language (Smil 2001a).

> Yesterday we began operating the large ammonia apparatus...and were able to keep its uninterrupted production for about five hours. During this whole time it...functioned correctly and it produced continuously liquid ammonia. Because of the lateness of the hour, and as we all were tired, we...stopped the production because nothing new could be learned from continuing the experiment.

Fritz Haber was a brilliant chemist. He also was an ardent patriot willing to do almost anything to serve the fatherland. During World War I, he led German efforts to utilize deadly chemicals such as phosgene and mustard gas as battlefield weapons. It has been reported that he personally supervised the use of poison gas at the battle of Ypres, in which more than 10,000 soldiers were killed in a matter of minutes. Many refer to him as the father of chemical warfare. His wife, also a chemist, was horrified by his activities, and some have speculated that this led her to commit suicide in the family living room by shooting herself in the heart. According to reports, Haber took little time to mourn. The day following her death, he left for the eastern front (Smil 2001b).

Carl Bosch, a gifted engineer, was given the task of taking what Haber had discovered in the laboratory and turning it into an industrial process. Bosch succeeded and soon Germany was producing ammonia in large quantities. Ammonia, or NH_3, is readily converted to ammonium nitrate or other compounds that can be used to fertilize the soil and greatly increase food production. But nitrogen has other uses. A line from an old nursery rhyme captures the essence of this element: "When she was good, she was very, very good, but when she was bad, she was horrid." In one of its good personas, as fertilizer, nitrogen can turn the landscape green and feed the hungry. At the time, however, Haber and his employers were not really concerned with the good potential of nitrogen. Their main interest was in preserving Germany's capacity to wage war; and nitrogen, in one of its "horrid" personas, is a strategic war material. Instead of making fertilizer, one can just as readily convert NH_3 to NO_3, which, when dissolved in water, makes nitric acid, or HNO_3. With HNO_3 as the starting point, one can create a veritable witch's brew of explosives, and that is what happened. Before it was used to feed millions of hungry people in the modern world, Haber's discovery first helped unleash the horrors of modern warfare, making the 20th century the bloodiest in history.

Fritz Haber has been praised by some and scorned by others, but one thing is undeniable. The successful experiment he carried out on July 2, 1909, was a watershed event in human history. His discovery of how to take nitrogen out of the air and turn it into fertilizer was a scientific and historic milestone for which he was awarded the Nobel Prize in 1918. The significance of Haber's work can hardly be exaggerated. By discovering how to convert nitrogen and hydrogen into ammonia (NH_3) under laboratory and ultimately factory conditions, Haber made it possible to produce unlimited quantities of nitrogen to fertilize crops. If not for this timely discovery, the dire prophecies of worldwide starvation in the 20th century probably would have been realized. When Crookes issued his warning in 1898, there were about 1.5 billon people in the world. By 2018, only 120 years later, the world's population had reached 7.6 billion; and nearly half of the people born during the 20th century were alive only because of synthetic nitrogen produced by the Haber-Bosch process. Demographers predict that billions more new people will be added during the 21st century, and these newcomers will be even more dependent on industrial nitrogen.

The age of industrial agriculture was born on July 2, 1909, but it did not truly come of age until 1947, when the United States government began converting wartime munitions plants into fertilizer factories, starting with a sprawling facility at Muscle Shoals, Alabama. The United States military had used a lot of explosives in World War II, and ammonium nitrate was the main ingredient. America built its first wartime ammonia plant in 1941, and by 1945 the nation had constructed ten additional plants that were turning out nearly 900,000 tons of ammonium nitrate each year. When hostilities ended, the nation found itself with a large surplus of this material. What should be done with all of the stockpiled nitrate and with the infrastructure that had been created to produce it? The United States Department of Agriculture came up with a solution: Use the left-over ammonium nitrate and the surplus production capacity to fertilize the nation's food crops.

The large amounts of cheap nitrogen fertilizer that became available in the last half of the 20th century had profound impacts on human societies. The biggest impact was on population. In 1900, the world had a population of 1.6 billion; as of 2018, it had grown to 7.6 billion, an increase of six billion in a little over a century. It has been estimated that the total weight of the human race is now about 340 billion kilograms, making our biomass at least 100 times greater than any other large species that has ever walked the Earth (Smil 2001b). This spectacular increase in human numbers and in human biomass during the last century would not have been possible without the Haber-Bosch process. Shortly after the end of World War I, Germany began converting its munitions plants to produce nitrogen fertilizer, and by 1920, the Haber-Bosch process accounted for nearly one fifth of the nitrogen fertilizer used worldwide. During the next 80 years, world production doubled nearly ten times, with annual usage increasing from 150,000 tons in 1920 to more than 85 million tons in 2000 (Smil 2001b; IFIA 1998).

Synthetic nitrogen fertilizer has become crucial to modern agriculture and to the human food supply because of the profound effect this element has on crop production. Studies have shown that, all else being equal, doubling the amount of nitrogen taken up by such crops as corn or rice will double the grain yield. If a well-managed rice crop takes up 50 kilograms of nitrogen per hectare, it can yield about 3,000 kilograms of grain per hectare; if nitrogen uptake is doubled to 100 kilograms per hectare, grain production will double to 6,000 kilograms per hectare (Crossman and Anderson 1992).

In 1900, the world's farmers were feeding about two people per hectare of farmland. At that time, most people in the world were eating a near-vegetarian diet, with only about ten percent of the grain harvest fed to animals. Let us now fast forward to 2018, when the world's farmers were feeding 7.6 billion people while cultivating 1.87 billion hectares of land (USGS 2017). Dividing 7.6 billion people by 1.87 billion hectares (7.6 /1.87) yields 4.1, or about four people per hectare. Using the best available data, the average carrying capacity of the world's farmland doubled from two people per hectare in 1900 to four people per hectare in 2018. But something else also happened. Not only were six billion more people alive in 2018 than in 1900; they also were eating a diet much higher in animal protein. In 2018, about half of the world's grain harvest was fed to animals, compared to just ten percent in 1900.

Two things made the huge population growth of the 20th century possible. One was the production of synthetic nitrogen, which helped double the amount of food produced on each hectare of land. At the same time, the amount of cultivated land also doubled, going from 850 million hectares in 1900 to 1,870 million hectares in 2017. Doubling productivity per hectare and doubling the amount of cultivated land enabled the world's farmers to feed a population five times as large a diet much higher in protein. But how was it possible for the world's farmland to more than double in little more than a century?

Emptying Out the Heartlands

At the dawn of the 20th century, there were more than 20 million horses on America's farms. Although possessing great power, horses are easily domesticated and trained, and they can do the work of seven or eight humans. The historical importance of horses is evidenced by the fact the amount of work a modern machine can do is still measured in units of horsepower. The typical American farmer in the early 20th century kept about four horses, the work equivalent of about 30 humans. These powerful, plodding beasts did most of the heavy work, but not for free. Animal labor on farms came at a steep price; not only did they require a lot of care, but millions of acres nationwide were set aside to feed them. In those days, horses and mules ate about 25 percent of the food produced, yet they were not able to do all of the work. Humans still had to perform a number of tasks, such as harvesting and processing corn, picking cotton, milking cows, and caring for all those animals. In the early decades of the 20th century, farming required a lot of muscle power from both animals and people.

Henry Ford changed all of that when, in 1921, he began selling a small tractor, the Fordson F, for $625. This tractor was rated at 20 horsepower, meaning it could do the work equivalent of 20 horses, or about 150 humans. It is no surprise that so many farmers found the economics of this compelling. Ford's tractors sold well and other companies soon entered the business. In 1922, the power takeoff was invented. This is a deceptively simple device, just a rigid, steel shaft with raised surfaces that make it fit like a key into another piece of machinery. The power takeoff enabled the tractor's engine to transfer power to a planter, mower, corn picker, or other implement being pulled behind. Engineers soon designed all kinds of farm equipment that would work off of a power takeoff. The results were revolutionary. For nearly 10,000 years, humans and animals had supplied all of the labor to plant, till and harvest crops. But this ended in just a few decades. By the middle of the 20th century, tractors and tractor-driven equipment were performing the heavy farm labor once done by humans and horses (Gray 1954). There are a few exceptions to this. For example, the farmers of India still rely heavily on animal power.

The advent of tractors and the power takeoff made it possible for each farmer to cultivate much more land. Replacing four horses with a tractor in the 1920s was roughly equivalent to acquiring more than 20 additional horses or nearly 150 additional farm laborers. Because of tractors and other mechanized equipment, farmers were able to plow up more than one billion hectares of new farmland during the 20th century, more than doubling the world's agricultural land base. Shifting to tractor power had another benefit. In places such as North America and Europe, the 20 to 25 percent of farmland that once had fed horses and mules could instead be used to grow food for human consumption.

Food in the Future

Population experts predict that there will be more than eleven billion people living on Earth by 2100. Considering this, two fundamental questions need to be addressed. First, will the world be able to feed that many people over the next few generations? The second question is this: If human population stabilizes at around eleven to twelve billion in 2100 as predicted, how long can the Earth support that many people — a hundred years, a thousand years, many thousands of years? The answer to the first question is rather straightforward. Using current agricultural technology, the world has more than enough arable land to feed twelve billion people or more. As stated earlier, the average carrying capacity of farmland is now about four people per hectare, and 1.87 billion hectares are currently being farmed worldwide. If productivity remains the same, the world can feed eleven billion people in 2100 by increasing the amount of farmland by about one billion hectares, and we have more than enough available land to do so.

Dutch scientists have estimated that there are about 3.42 billion hectares of arable land in the world (Buringh 1977). This estimate, based on the 222 soil regions of the FAO/UNESCO World Soil Map, is in close agreement with an earlier assessment made by the President's Science Advisory Committee (1967). Fortunately, we have an equally reliable estimate of the amount of land currently being cultivated. In 2017, a study led by the United States Geological Survey determined that worldwide, about 1.87 billion hectares of land, or about 55 percent of the total available, were being farmed. If the amount of cultivated land were increased to three billion hectares in order to feed twelve billion people in 2100, we then would be using around 90 percent of the world's arable land.

This rather simple assessment agrees with more comprehensive analyses such as those by Buringh (1977) and Eswaran et al. (1999). According to Eswaran et al. (1999), "Anticipated advances in biotechnology and sustainable land management, in combination with the availability of high quality land, suggest a level of food production that will sustain twice the current global population." Buringh (1977) also maintained that the world's farmers would have little problem feeding a population of twelve billion people. Vaclav Smil, perhaps the most knowledgeable individual among us on this subject, concurred, writing in 2001, "In global terms, the availability of farmland is not a limiting factor in the quest for decent nutrition during the next two generations."

The world has enough land, fossil fuels, minable phosphate, and other agricultural inputs to feed a population of eleven to twelve billion for the short term, but longer term the prognosis is less certain. As articulated by Smil (2001a), agriculture as now practiced "is increasingly dependent on uncertain inputs of fossil fuels, and hence inherently unstable on a civilizational timescale (10^3 years)." This chapter began with Sir William Crookes' 1898 address to the British Association for the Advancement of Science in which he offered the following grim assessment: "England and all civilized nations stand in deadly peril of not having enough to eat...we are drawing on the earth's capital, and our drafts will not perpetually be honored." The imminent catastrophe that Crookes and others so feared was avoided. By discovering how to remove nitrogen gas from the air and convert it into nitrogen fertilizer, Fritz Haber gave us a gift — a second chance.

Unfortunately, we have squandered that second chance and are once again facing a crisis. Humanity must begin taking steps to bring its numbers into line with the planet's innate carrying capacity. It is imperative that we determine how many people Earth can support in perpetuity without degrading or destroying the soil that feeds us and makes human civilization possible. How we do that is well beyond the scope of this book. But, as a very minimum, scientists, government leaders, and ordinary people the world over must learn to "see" soil as part of the environment, come to appreciate how essential it is to life, and understand why preserving it has become one of humanity's most critical issues. If this book helps to achieve that, it will have been worth it.

REFERENCES

Aitken, J. (1881). Dust, fogs, and clouds. *Nature* 23(591), 384–85.

Akimtzev, V.V. (1932). Historical soils of the Kamenetz-Podolsk Fortress. *Proc. Second Int. Cong. Soil Sci.* 5, 132–40.

Allaart, J.H. (1976). The pre-3760 m.y. old supracrustal rocks of the Isua area, central west Greenland, and the occurrence of quartz-banded ironstones. Pp. 177–189 in Windley, B.F. (ed.). *The early history of the Earth*. Wiley-Interscience, London.

Allen, O.E. (1983). *Atmosphere*. Time-Life Books, Alexandria, VA.

Aulie, R.P. (1970). Boussingault and the nitrogen cycle. *Amer. Phil. Soc.* 114(6), 435-479.

Barber, J. (2003). Photosystem II: the engine of life. *Q. Rev. Biophysics* 36(1), 71–89.

Bentley, W.A. (1904). Studies of raindrops and raindrop phenomena. *Monthly Weather Review* 32, 450–456.

Berner, R.A. (1997). The rise of plants and their effect on weathering and atmospheric CO_2. *Science* 276, 544–545.

Black, C.A. (1957). *Soil-plant relationships*. John Wiley and Sons, New York.

Blanchard, D.C. (1966). *From raindrops to volcanoes: Adventures with sea surface meteorology*. Anchor Books, Doubleday and Company, Garden City, NY.

Blanchard, D.C. (1972). Bentley and Lenard: Pioneers in cloud physics. *Amer. Sci.* 60, 746–749.

Blanchard, D.C. (1998). *The snowflake man: A biography of Wilson A. Bentley*. McDonald and Woodward Publishing Co., Granville, OH.

Bohn, H.L., McNeal, B.L., and O'Connor, G.A. (1979). *Soil chemistry*. John Wiley and Sons, NY.

Braidwood, R.J. (1964). *Prehistoric men*. Chicago Natural History Museum, Chicago.

Brewer, R. (1964). *Fabric and mineral analysis of soils*. John Wiley and Sons, New York.

Budiansky, S. (1999). *The covenant of the wild: Why animals chose domestication*. Yale University Press, New Haven, CT.

Buol, S.W., and Hole, F.D. (1961). Clay skin genesis in Wisconsin soils. *Soil Sci. Soc. Am. Proc.* 25, 377–379.

Burbidge, E.M., Burbidge, G.R., Fowler, W.A. and Hoyle, F. (1957). Synthesis of the elements in stars. *Reviews of Modern Physics* 29(4), 547.

Buringh, P. (1977). Food production potential of the world. *World Development* 5, 477–485.

Cato, M.P. *De agri cultura (On agriculture)*. Translated by W.D. Hooper. (1934). Harvard University Press, Cambridge, MA.

Chiarenzelli, J.R., Donaldson, J.A., and Best, M. (1983). Sedimentology and stratigraphy of the Thelon Formation and the sub-Thelon regolith. *Papers of the Geological Survey of Canada* 83–1A, 443–445.

Christian, D. (2005). *Maps of time: An introduction to big history*. University of California Press, Berkeley, CA.

Clarke, F.W. (1924). *Data of geochemistry*. USGS. Reston, VA.

Cohen, N.C. (1977). *The food crisis in prehistory: Overpopulation and the origins of agriculture*. Yale University Press, New Haven, CT.

Columella, L.J. *Res rusticae (Country matters)*. Translated by Ash, H.B. (1934). Harvard University Press, Cambridge, MA.

Cooper, A.W. (1960). An example of the microclimate in soil genesis. *Soil. Sci.* 90, 109–120.

Correns, D.W. (1949). *Introduction to mineralogy*. Springer-Verlag, Berlin.

Coulier, J-P. (1875). Note sur une nouvelle propriété de l'air. *Jour. de Pharmacie et de Chimie*. Series 4, 22(1), 165–172 and 22(2), 254–255.

Crookes, W. (1899). *The wheat problem*. Chemical News Office, London.

Crossman, P., and Anderson, J.R. (1992). *Resources and global food prospects: supply and demand for cereals to 2030*. World Bank, Washington, DC.

Daniels, R.B., Kleiss, H.J, Buol, S.W, Byrd, H.J. and Phillips, J.A. (1984). *Soil systems in North Carolina*. North Carolina Agricultural Research Service, Raleigh, NC.

Darwin, C. (1881). *The formation of vegetable mould through the action of worms: with observations on their habits*. John Murray, London.

DeGaetano, A.T. (1994). *Daily evapotranspiration and soil moisture estimates for the northeastern United States*. Northeast Regional Climate Center, Cornell University, Ithaca, NY.

Dell'Amore, C. (2009). Amazon's low salt content keeps carbon emissions at bay. *National Geographic News*. Retrieved from http//www.nationalgeographic.com

Diamond, J. (1987). The worst mistake in the history of the human race. *Discover*, 64–66.

Diemer, J.A., and Bobyarchick, A.R. (2005). Coastal plain. *The North Carolina Atlas Revisited*. University of North Carolina, Chapel Hill.

Editors, Time-Life Books. (1985). *Grasslands and tundra*. Time-Life Books, Alexandria, VA.

Eglinton, G., Maxwell, J.R., and Pillinger, C.T. (1972). The carbon chemistry of the moon. *Sci. Amer.* 227(4), 81–90.

Eldridge, N., and Gould, S.J. (1972). Punctuated equilibria: An alternative to phyletic gradualism. Pp. 82–115 in Schopf, T.J.M. (ed.). *Models in Paleobiology*. Freeman Cooper, San Francisco.

Emlen, J.M. (1966). The role of time and energy in food preference. *Amer. Naturalist* 100, 611–617.

Eswaran, H., Beinroth, F., and Reich, P. (1999). Global land resources and population supporting capacity. *Amer. Jour. Alternative Agriculture* 14, 129–136.

Falini, F. (1965). On the formation of coal deposits of lacustrine origin. *Bull. Geol. Soc. Amer.* 76, 1317–1346.

Falkowski, P.G. (2015). *Life's engines: How microbes made Earth habitable*. Princeton University Press, Princeton, NJ.

Fallou, F.A. (1862). *Pedology or general and special soil science*. G. Schonfeld, Dresden.

Farb, P. (1963). *Face of North America: The natural history of a continent*. Harper & Row, New York.

Farb, P. (1978). *Humankind*. Houghton-Mifflin, Englewood Cliffs, NJ.

Farnsworth, R.K., and Thompson, E.S. (1982). *Mean monthly, seasonal, and annual pan evaporation for the United States*. National Weather Service, Washington, DC.

Ferwerda, J.A., LaFlamme, K.J., Kalloch, N.R. and Rourke, R.V. (1997). *The soils of Maine.* Maine Agricultural and Forest Experiment Station and USDA-Natural Resources Conservation Service, Washington, DC

Floudas, D., et al. (2012). The Paleozoic origin of enzymatic lignin decomposition reconstructed from 31 fungal genomes. *Science* 336, 1715–1719.

Gagnon, J.A. (2001). *Soil survey of Hyde County, North Carolina.* Natural Resources Conservation Service, Washington, DC.

Gambell, A.W., and Fisher, D.W. (1966). *Chemical composition of rainfall: eastern North Carolina and southeastern Virginia.* U.S. Govt. Printing Office, Washington, DC.

Gillespie, W.H., Rothwell, G.W., and Scheckler, S.E. (1981). The earliest seeds. *Nature* 293, 462–464.

Goldfarb, B. (2018). *Eager: The Surprising secret life of beavers and why they matter.* Chelsea Green Publishing, White River Junction, VT.

Goldich, S.S. (1938). A study in rock weathering. *Jour. Geology* 46, 17–58.

Grandstaff, D.E., Edelman, M.J., Foster, R.W., Zbinden, E, and Kimberley, M.M. (1986). Chemistry and mineralogy of Precambrian paleosols at the base of the Dominion and Pongola Groups. *Precambrian Research* 32, 97–131.

Gray, R.B. (1954). *The agricultural tractor: 1855–1950.* American Society of Agricultural Engineers, St. Joseph, MI.

Grim, R.E. (1953). *Clay mineralogy.* McGraw-Hill, New York.

Harris, R.W. (1992). Root–shoot ratios. *Jour. Arboriculture* 18(1), 39–42.

Harrison, J.B. (1933). *The katamorphism of igneous rocks under humid tropical conditions.* Imperial Bureau of Soil Science, Harpenden, England.

Havlin, J.L., Tisdale, S.L, Nelson, W.L., and Beaton, J.D. (2016). *Soil fertility and fertilizers: An introduction to nutrient management (8th edition).* Pearson India, Delhi.

Hazen, R.M. (2012). *The story of Earth: The first 4.5 billion years, from stardust to living planet.* Penguin Books, New York.

Hazen, R.M., and Trefil, J. (2009). *Science matters: Achieving scientific literacy.* Anchor Books, New York.

Hellriegel, H., and Wilfarth, H. (1891). Recherches sur l'alimentation azotées des graminées et des legumineuses. *Ann. Sci. Agron.* 7, 84–175 and 189–352.

Hilgard, E.W. (1914). *Soils.* The MacMillan Company, New York.

Hodson, M.J., White, P.J., Mead, A., and Broadley, M.R. (2005). Phylogenetic variation in the silicon composition of plants. *Ann. Bot.* 96(6), 1027–1046.

Hole, F.D., and Bidwell, O.W. (1989). Proposal for a national soil of the United States. *Soil Survey Horizons* 30(3), 77–79.

Hudson, B. (1984). *Soil survey of Cumberland and Hoke counties, North Carolina.* USDA-Soil Conservation Service, Washington, DC.

Hudson, B. (1992). Soil survey as paradigm-based science. *Soil Sci. Soc. Amer.* 56(3), 836–841.

Hudson, B. (1994). Soil organic matter and available water capacity. *Jour. Soil and Water Conservation* 49(2), 189–194.

Huggins, M.L., and Sun, K.H. (1946). Energy additivity in oxygen-containing crystals and glasses. *Jour. Physical Chem.* 50, 319–328 and 438–443.

International Fertilizer Industry Association. (1998). *World nitrogen fertilizer consumption.* IFA, Paris. Retrieved from http://www.fertilizer.org.

Jacobs, W. (1937). Preliminary report on a study of atmospheric chlorides. *Monthly Weather Rev.* 65, 147–151.

Jenny, H. (1941). *Factors of soil formation.* McGraw-Hill, New York.

Jenny, H., Leonard. C.D. (1934). Functional relationships between soil properties and rainfall. *Soil Sci.* 38, 363–381.

Junge, C.E. (1958). The concentration of chloride, sodium, potassium, calcium, and sulfate in rain water over the United States. *Jour. Meteorology* 15(5), 417–425.

Junge, C.E., and Gustafson, P.E. (1957). On the distribution of sea salt over the United States and its removal by precipitation. *Tellus* 9, 164–173.

Kant, I. (1781). *Critique of pure reason.* Translation published 1999 by Cambridge University Press, Cambridge, UK.

Kaspari, M., Yanoviak, S.P., and Clay, N.A. (2009). Sodium shortage as a constraint on the carbon cycle in an inland tropical rain forest. *Proc. Natl. Acad. Sci.* 106(46), 19405–19409.

Keller, W.D. (1957). *Principles of chemical weathering.* Lucas Bros. Publishing, Columbia, MO.

Kellogg, C.E. (1950). Soil. *Scientific Amer.* 183(1), 30–39.

Kennedy, J.R. (2006). Salt licks. *NCPedia, the online encyclopedia of North Carolina.* Retrieved from http://www.ncpedia.org.

Khalifa, E.M., and Buol, S.W. (1968). Studies of clay skins in a cecil (typic hapludults) soil. I. composition and genesis. *Soil Sci. Soc. Am. Proc.* 32, 857–861.

Kidston, R., and Lang, W.H. (1921). On old red sandstone plants showing structure from the rynie chert bed, Aberdeenshire. Part V. The thallophyta occurring in the peat bed, the succession of plants through a vertical section of the bed and the conditions of accumulation and preservation of the deposit. *Transactions of the Royal Society of Edinburgh* 52, 855–902.

King, F.H. (1927). *Farmers of forty centuries.* Harcourt, Brace & Company, New York.

Kladivko, E.J. (1993). *Earthworms and crop management.* Purdue University Cooperative Extension Service, West Lafayette, IN.

Kleier, C. (2017). *Plant science: An introduction to botany.* The Teaching Company. Chantilly, VA.

Klingebiel, A.A. (1966). Costs and returns of soil surveys. *Soil Conservation* 32, 3–6.

Knoll, M.A., and James, W.C. (1987). Effect of the advent and diversification of vascular plants on mineral weathering through geologic time. *Geology* 15, 1099–1102.

Kurlansky, M. (2002). *Salt: A world history.* Penguin Books, New York.

Langmuir, I., and Blodgett, K.B. (1946). *A mathematical investigation of water droplet trajectories.* Collected works of I. Langmuir, vol. 10, 348–393. Pergamon, London.

Laws, J.O., and Parsons, D.A.(1943). The relation of rain-drop size to intensity. *Trans. Amer. Geophys. Union* 24, 452–460.

Le Chatelier, H. (1887). De l'action de la chaleur sur les argiles. *Bull. Soc. Franc. Mineral* 10, 204–2011.

Lee, R.B., and Devore, I. (1968). *Man the hunter.* Aldine, Chicago.

Legge, A.J., and Rowley-Conwy, P.A. (1987). Gazelle killing in Stone Age Syria. *Sci. Amer.* 257, 88–95.

Lenard, P. (1904). Uber regen. *Meteorol. Zeit.* 21, 248–62.

Lenton, T.M., Crouch, M., Johnson, M., Pires, N., and Dolan, L. (2012). First plants cooled the Ordovician. *Nature Geoscience* 5, 86–89.

Livingstone, D.A. (1963). *Chemical composition of river water and lakes. Data of geochemistry.* U.S. Geological Survey, Reston, VA.

Losche, C.K. (1967). *Soil genesis and forest growth on steeply sloping landscapes of the Southern Appalachians.* Ph.D. thesis, North Carolina State University, Raleigh.

Low, A.J. (1955). Improvements in the structural state of soils under leys. *Jour. Soil Sci.* 6, 179–199.

Lowe, D.R. (1983). Restricted shallow-water sedimentation of early Archaean stromatolitic and evaporitic strata of the Strelley Pool chert, Pilbara block, Western Australia. *Precambrian Research* 19, 239–283.

MacArthur, R.H., and Pianka, E.R. 1966. On the optimal use of a patchy environment. *The Amer. Naturalist* 100, 603.

Malcom, N. (2012). *Thomas Hobbes: Leviathan*. Oxford University Press, Oxford.

Mandelbrot, B. (1967). How long is the coast of Britain? Statistical self-similarity and fractal dimension. *Science* 156(3775), 636–638.

Martin, S.J., Funch, R.R., Hanson, P.R. and Yoo, E-H. (2018). A vast 4,000-yr-old spatial pattern of termite mounds. *Current Biol.* 28(22), 1292–1293.

Mason, B. (1952). *Principles of geochemistry*. John Wiley and Sons, New York.

McCaleb, S.B. (1959). The genesis of red-yellow podzolic soils. *Soil Sci. Soc. Am. Proc.* 23, 164–168.

McPherson, J.G. (1979). Calcrete (caliche) paleosols in fluvial redbeds of the Aztec siltstone (Upper Devonian), southern Victoria Land, Antarctica. *Sedimentary Geology* 22, 319–320.

Merrill, G.P. (1897). *Rocks, rock-weathering and soils*. MacMillan Co., New York.

Mount, J.L. (1975). *The food and health of Western man*. John Wiley and Sons, New York.

National Agricultural Statistics Service. (2015). *Crop production 2014 summary*. USDA, Washington, DC.

Novaes, E., Kirst, M., Chiang, V., Winter-Sederoff, H., and Sederoff, R. 2010. Lignin and biomass: A negative correlation for wood formation and lignin content in trees. *Plant Physiol.* 154(2), 555–561.

Paoletti, M.G., Pimentel, D., Stinner, B.R., and D. Stinner, D. (1992). Agroecosystem biodiversity: matching production and conservation biology. *Agriculture, Ecosystems, and Environment* 40, 3–23.

Parker, R.B. (1984). *Inscrutable Earth: Explorations into the science of Earth*. Scribner, New York.

Parker, R.L. (1967). *Composition of the earth's crust, 6th Edition*. U.S. Geological Survey, Washington, DC.

Pauling, L. (1929). The principles determining the structure of complex ionic crystals. *Jour. Amer. Chem. Soc.* 51(4), 1010–1026.

Pauling, L. (1960). *The nature of the chemical bond and the structure of molecules and crystals: an introduction to modern structural chemistry (3rd edition)*. Cornell University. Press, Ithaca, NY.

Penman, H.L. (1941). Laboratory experiments on evaporation from fallow soil. *Jour. Agric. Sci.* 31, 454–465.

Percival, C.J. (1986). Paleosols containing an albic horizon: examples from the Upper Carboniferous of northern England. Pp. 87–11 in Wright, P.V. (ed.). *Paleosols: Their recognition and interpretation*. Blackwell, Oxford.

Persaud, C. (1974). *Potential evapotranspiration in different climatic regions of Guyana*. Master of science thesis, McGill University, Quebec, Canada.

Philbrick, N. (2006). *Mayflower: A story of courage, community, and war*. Viking, the Penguin Group, New York.

Polynov, B.B. (1927). *Contributions of Russian scientists to paleopedology*. USSR Academy of Sciences, Leningrad.

Polynov, B.B. (1937). *The cycle of weathering*. Thomas Murby and Co., London.

President's Science Advisory Council. (1967). *The world food problem*. The White House. Washington, DC.

Pringle, L. (1985). *Rivers and lakes*. Time-Life Books, Alexandria, VA.

Rackham, O. (1990). *Trees and woodland in the British landscape*. J.M. Dent and Sons, London.

Raven, J.A. (1983). The transport and function of silica in plants. *Biol. Rev.* 58(2), 179–207.

Raven, J.A., and Edwards, D. (2001). Roots: Evolutionary origins and biogeochemical significance. *Jour. Experimental Botany* 52, 381–401.

Raymond, J., Siefert, J.L., Staples, C.R. and Blankenship, R.E. (2004). The natural history of nitrogen fixation. *Mol. Biol. Evol.* 21, 541–554.

Retallack, G.J. (1986). Reappraisal of a 2200-ma-old paleosol from near Waterval Onder, South Africa. *Precambrian Research* 32, 195–252.

Retallack, G.J. (1990). *Soils of the past: An introduction to paleopedology*. Unwin Hyman, Boston.

Robinson, J.M. (1990). Lignin, land plants, and fungi: Biological evolution affecting phanerozoic oxygen balance. *Geol.* 18(7), 607.

Ronov, A.B., and Yaroshevsky, A.A. (1972). Earth's crust geochemistry. Pp. 234–254 in Fairbridge, F.W. (ed.), *Encyclopedia of geochemistry and environmental sciences*. Van Nostrand Reinhold, New York.

Ross, C.S. (1927). The mineralogy of clays. *First International Congress of Soil Science* 4, 555-561.

Ross, C.S., and Hendricks, S.B. (1945). *Minerals of the montmorillonite group*. U.S. Geol. Survey, Washington, DC.

Ross, C.S., and Kerr, P.F. (1931). *The kaolin minerals*. U.S. Geol. Survey, Washington, DC.

Russell. E.J. (1911). *Lessons on soil*. Cambridge University Press, Cambridge, UK.

Russell, E.J. (1957). *The world of the soil.* Collins Clear-Type Press, London and Glasgow.

Rye, R., and Holland, H.D. (1998). Paleosols and the evolution of atmospheric oxygen: A critical review. *Science* 298, 621–672.

Rynasiewicz, J. (1945). Soil aggregation and onion yields. *Soil Sci.* 60, 387–395.

Sagan, C. (1980). *Cosmos.* Random House, New York.

Sainju, U.M., Allen, B.L., Lenssen, A.W., and Ghimire, R.P. (2017). Root biomass, root/shoot ratio, and soil water content under perennial grasses with different nitrogen rates. *Field Crops Research* 210, 183–191.

Shaw, C.F. (1927). The normal moisture capacity of soils. *Soil Sci.* 23, 303–317.

Shiklomanov, I. (1993). World fresh water resources. In Gleick, P.H. (ed.). *Water in crisis: a guide to the world's fresh water resources.* Oxford University Press, New York.

Shul'gin, A.M. (1957). *The temperature regime of soils.* Translated by Gourevich, A. Israel Program for Sci. Trans. Cat. No. 1357, 1965.

Smil, V. (1997). *Cycles of life: civilization and the biosphere.* W.H. Freeman and Company, New York.

Smil, V. (2001a). *Enriching the earth: Fritz Haber, Carl Bosch, and the transformation of world food production.* The MIT Press, Cambridge, MA.

Smil, V. (2001b). *Feeding the world: A challenge for the twenty-first century.* The MIT Press, Cambridge, MA.

Smith, G.S., Middleton, K.R., and Edmonds, A.S. (1978). A classification of pasture and fodder plants according to their ability to translocate sodium from their roots into aerial parts. *New Zealand Jour. Experimental Agric.* 6, 183.

Soil Survey Staff. (2015). *Illustrated guide to soil taxonomy, version 2.* Natural Resources Conservation Service, Washington, DC.

Stanworth, C.W., and Badham, J.P.N. (1984). Lower proterozoic red beds, evaporites and secondary sedimentary uranium deposits from east arm, great slave lake, Canada. *Jour. Geol. Soc.,* London 141, 235–242.

Steinbeck, J. (1939). *The grapes of wrath.* Viking Press, New York.

Stuhlman, O. (1932). The mechanics of effervescence. *Physics* 2, 457.

Subbarao, G.V., Ito, O., Berry, W.L., and Wheeler, R.M. (2003). Sodium—a functional plant nutrient. *Critical Reviews in Plant Sciences* 22(5), 391–416.

Sung, Ying-hsing. (1637). *The creations of nature and man.* Translated by Sun, E.Z., and Sun, S.C. (1966). Pennsylvania State University Press, University Park, PA.

Sverdrup, H.V., Johnson, M.W., and Fleming, R.H. (1946). *The oceans*. Prentice-Hall, New York.

Tanaka, Y. (1998). The cyclical sensibility of Edo-period Japan. *Japan Echo* 25(2), 12–16.

Testo, W., and M. Sundue, M. (2016). A 4000-species dataset provides new insight into the evolution of ferns. *Molecular Phylogenetics and Evolution* 105, 200–211.

Theophrastus. *Enquiry into plants, volume 1: books 1-5*. Trans. by Hort, A.F. (1916). Loeb Classical Library 70. Harvard University Press, Cambridge, MA.

Thomas, E.M. (1959). *The harmless people*. Knopf, New York.

Thorpe, J., Cady, J.G., and Gamble, E.E. (1959). Genesis of Miami silt loam. *Soil Sci. Soc. Am. Proc.* 23, 156–161.

USFS. (2014). *U.S. forest resource facts and historical trends*. U.S. Forest Service, Washington, DC.

USGS. (2017). *New map of worldwide croplands supports food and water security*. U.S. Geological Survey, Washington, DC. Retrieved from http://www.usgs.gov.

van Es, H. (2017). A new definition of soil. *CSA News* 62(10), 20–21.

van Soest, P.J. (1982). *Nutritional ecology of the ruminant*. Cornell University Press, Ithaca, NY.

Veihmeyer, F.J. (1927). Some factors affecting the irrigation requirements of deciduous orchards. *Hilgardia* 2, 125–291.

Veihmeyer, F.J., and Hendrickson, A.H. (1931). The moisture equivalent as a measure of field capacity in soils. *Soil Sci.* 32, 181–193.

Vinogradov, A.P. (1956). Distribution patterns of chemical elements in the earth's crust. *Geokhimiya (Geochemistry)* 1, 1–43.

Watanabe, Y., Martini, J.E., and Ohmoto, H. (2000). Geochemical evidence for terrestrial ecosystems 2.6 billion years ago. *Nature* 408, 574–578.

Weaver, J.E. (1926). *Root development of field crops*. McGraw-Hill, New York.

Weaver, J.E., and Zink, E. (1946). Annual increase of underground materials in three range grasses. *Ecology* 27, 115–127.

Wharton, C.H. (2001). *Metabolic man: Ten thousand years from Eden*. WinMark Publishing, Orlando, FL.

Williams, D.L. (2014). *Rousseau's social contract: An introduction*. Cambridge University Press, Cambridge, UK.

Wolf, E.R. (1966). *Peasants*. Prentice-Hall, Englewood Cliffs, NJ.

Young, G.M., and Long, D.G.F. (1976). Ice wedge casts from the Huronian Ramsey Lake formation (2300. m. y. old), near Espanola, northern Canada. *Paleogeography, Paleoclimatology, Paleoecology* 19, 191–200.

Zubrin, R. 2014. The Pacific's salmon are back—thank human ingenuity. *National Review*, April 22. Retrieved from http://www.nationalreview.com.

INDEX

genes, 94, 119, 121, 203

glacial, 31, 46, 51, 127, 136, 138, 142, 143

glaciation, 15, 139, 160, 166, 179, 182, 183, 197

glucose, 10, 28, 101, 119, 120, 206

granite, 9, 12, 19, 26-28, 31, 42, 47, 78, 99, 114, 127, 128, 138

granitic, 9, 27, 30, 124, 140-142, 146

gravity, 5, 8, 21-24, 31, 33, 34, 39, 53, 96, 108, 110-112, 116, 168, 169

green clay, 14, 180, 182

Grim, Ralph, 60, 78, 194, 195, 201, 210, 224

Guyana, 72, 74, 146-148

H

Haber, Fritz, 219, 220, 225

Hadean, 24

Hazen, Robert, 7, 8, 21, 30, 60

helium, 21, 30, 33, 37

Hellriegel, Hermann, 204

hemoglobin, 203, 205

Hilgard, Eugene, 15, 124, 183, 200

Histosols, 14, 16, 70, 127, 134, 165-170, 179, 181, 184, 191, 198

horsepower, 222, 223

horses, 91, 159, 164, 216, 222, 223

humification, 190, 191, 193

humus, 190, 191, 193, 198

hunter-gatherers, 90, 207-211, 215, 216, 219

hydration, 46

hydrogen, 10, 21-23, 27, 30, 33, 39, 47, 60, 61, 63, 71, 82, 97, 101, 110, 112, 202, 204, 220

hydrologic cycle, 187-189

hydrolysis, 48, 49, 179

I

ice age, 73, 105, 166, 168

igneous, 31, 32, 42, 47, 48, 78, 126, 130

illite, 79, 194, 195

illuvial, 145, 192, 199

Inceptisols, 13-15, 70, 146, 168-170, 179-183, 196

ionic, 39, 40, 51, 57, 98, 99

iron, 3, 9, 14, 20-23, 26-29, 32, 33, 36, 38, 40, 41, 44, 45, 50-52, 68, 70, 71, 73-75, 77, 79, 80, 113, 126-128, 140, 146, 148-150, 154, 183, 191, 197-199

J

Jenny, Hans, 4, 11, 48, 51, 65, 66, 69-72, 123, 124, 129-131, 170, 183

K

Kamenetz, 51

Kant, Immanuel, 6

kaolinite, 60-63, 68, 117, 140, 146, 148, 150, 194

Keller, Walter, 45, 47, 48, 50, 51, 67, 71, 88, 135, 187

Kellogg, Charles, 4

Klingebiel, Albert, 177, 178

Kurlansky, Mark, 74, 88, 89

L

landform, 174, 175

landscape, 12, 16, 41, 69, 72, 104, 108, 124, 134, 137, 149, 150, 164-166, 168, 169, 171, 173-175, 183, 212, 220

laterite, 71, 72

lateritic, 67, 68, 71

leaching, 52, 65-68, 70, 72, 73, 88, 131, 133, 135-137, 139-142, 146-150, 192, 194, 198

legume, 205, 219

Lenard, Philipp, 110, 111

lignin, 10, 97, 104, 105, 107, 155-157, 184, 185

limestone, 12, 31, 32, 42, 81, 82, 124, 146

loess, 31, 46, 51, 52, 66, 67, 127, 129, 131, 140, 169, 170, 214

Printed in the United States
By Bookmasters